AI AWARENESS SERIES

AI in Human Resources

Iris Stonefield

Contents

Introduction

The human resources function stands at an inflection point. For decades, HR departments have wrestled with an inherent tension: how to manage the deeply human aspects of work—motivation, development, culture, and relationships—while simultaneously handling the crushing volume of administrative tasks, data processing, and compliance requirements that modern organizations demand. Today, artificial intelligence offers a resolution to this paradox, not by replacing the human element, but by amplifying HR's capacity to focus on what truly matters: people.

The transformation is already underway. Organizations worldwide are deploying AI systems that can screen thousands of resumes in minutes, predict which employees are at risk of leaving months before they resign, and deliver personalized learning experiences that adapt in real-time to individual needs. Chatbots handle routine employee queries around the clock, while sophisticated algorithms analyze communication patterns to gauge team morale and identify emerging workplace issues before they escalate. These aren't futuristic concepts—they're tools being used today by companies ranging from tech startups to traditional manufacturers.

Yet this technological revolution in HR extends far beyond simple automation. AI is fundamentally reshaping how organizations think about talent. Where hiring once relied heavily on gut instinct and subjective impressions, machine learning models now identify candidates whose skills and experiences predict success in specific roles. Where performance reviews happened annually with backward-looking assessments, continuous AI-powered feedback systems provide real-time insights that help employees grow. Where succession planning depended on informal observations and office politics, data-driven analytics now map talent pipelines with unprecedented objectivity.

This shift brings both tremendous opportunities and sobering responsibilities. The promise is compelling: reduced bias in hiring, more equitable performance evaluations, personalized development paths for every employee, and HR teams freed from administrative burden to focus on strategic initiatives. But the challenges are equally significant. How do we ensure AI systems don't perpetuate historical discrimination? How do we maintain the human touch in employee relations while leveraging algorithmic efficiency? How do we protect employee privacy while gathering the data necessary for AI to function effectively?

This book provides practical guidance for navigating these complexities. Rather than offering theoretical speculation about AI's potential, we focus on proven applications, real-world implementations, and concrete strategies for success. Each chapter builds systematically from foundational concepts to advanced applications, ensuring readers develop both technical understanding and strategic insight.

Whether you're an HR professional seeking to understand AI's implications for your field, a business leader evaluating AI investments, or a technology professional working to implement HR systems, this book provides the knowledge framework necessary for success. We explore not just the "what" and "how" of AI in HR, but crucially, the "why" and "when"—helping you make informed decisions about which technologies to adopt, how to implement them ethically, and when human judgment must take precedence over algorithmic recommendations.

The future of HR isn't about choosing between human insight and artificial intelligence. It's about orchestrating both in harmony, creating workplaces that are simultaneously more efficient and more human, more data-driven and more empathetic, more standardized and more personalized. This book shows you how to achieve that balance, transforming HR from a support function into a strategic driver of organizational success.

Chapter 1: Understanding AI in the HR Context

Definitions and Key Concepts

Let's examine the practical applications of AI in HR today. Automated resume screening eliminates the time-intensive manual review process while reducing unconscious bias in initial candidate selection. AI-powered candidate matching goes beyond keyword searches, analyzing skills, experience patterns, and cultural fit indicators to identify optimal matches. Employee performance analysis leverages data from multiple touchpoints to provide comprehensive insights into productivity and development needs. Meanwhile, AI chatbots serve as 24/7 HR assistants, handling routine inquiries about policies, benefits, and procedures, freeing up HR professionals for strategic work.

Benefits and Challenges of AI in HR

AI implementation brings significant advantages but also notable challenges. Efficiency gains through automation can reduce processing times from days to minutes, allowing HR teams to focus on strategic initiatives rather than administrative tasks. Data-driven decision making replaces gut instincts with evidence-based insights, leading to better hiring decisions and employee outcomes. However, algorithmic bias poses serious risks if AI systems perpetuate historical discrimination patterns. Privacy concerns arise from increased data collection and analysis. Additionally, organizational resistance to change can impede successful AI adoption, requiring careful change management strategies.

Legal and Ethical Considerations

Responsible AI implementation requires strict adherence to legal and ethical standards. Data protection compliance with regulations like GDPR and CCPA is non-negotiable when handling employee information. Transparency in AI decision-making processes builds trust and enables employees to understand how conclusions are reached. Ethical fairness demands continuous monitoring for discriminatory outcomes and regular algorithm auditing. Employee privacy protection goes beyond legal compliance to include ethical data use and clear communication about how personal information is processed. These considerations form the foundation of trustworthy AI systems in HR.

By automating routine tasks and providing data-driven insights, AI significantly enhances HR efficiency and decision-making quality. However, successful implementation requires organizations to proactively address the challenges we've discussed - from bias and privacy concerns to change resistance. The key to success lies in balancing technological advancement with ethical responsibility, ensuring that AI serves to enhance human potential rather than replace human judgment. Moving forward, HR professionals must become AI-literate to leverage these tools effectively while maintaining the human-centered approach that defines great HR practice.

Chapter 2: Data Readiness for HR AI

Human Resource Information Systems serve as the technological backbone of modern HR operations. HRIS software manages essential functions including employee data storage, payroll processing, and recruitment workflows, creating a centralized repository for all HR information. This centralization improves data accessibility and management efficiency across organizations. However, HRIS implementations vary significantly between organizations, affecting data consistency and availability. Understanding your organization's specific HRIS architecture is crucial before implementing AI solutions, as it determines what data sources you'll have access to and how easily they can be integrated.

Organizations face significant challenges when attempting to integrate disparate HR data sources for AI applications. Different data formats across systems complicate seamless integration, while inconsistent data standards lead to discrepancies and unreliable datasets. Legacy systems present particular obstacles, often hindering smooth data sharing due to outdated architectures and limited connectivity options. Additionally, the absence of real-time data integration capabilities delays accurate HR analytics and can impact AI model performance. These challenges require strategic planning and technical solutions to overcome effectively.

Successful data integration requires implementing several best practices to ensure seamless connectivity across HR systems. Standardized data formats are essential for consistent exchange between diverse platforms, reducing compatibility issues and improving integration reliability. Middleware solutions act as bridges connecting disparate systems, facilitating smooth data flow and communication. Establishing comprehensive data governance policies ensures quality, security, and compliance throughout integration processes. Leveraging APIs enables real-time data exchange between HR platforms and AI applications, providing the dynamic connectivity necessary for effective AI implementation and ongoing model performance.

Data quality issues represent fundamental obstacles to successful AI implementation in HR contexts. Common problems include missing values, duplicate records, inconsistencies across datasets, and outdated entries that compromise analysis accuracy. Proactive detection of these issues is essential for maintaining data integrity and reliability before feeding information into AI models. Early identification allows organizations to address problems systematically rather than reactively. The benefits of correcting data quality issues are substantial, directly improving AI model accuracy and performance while ensuring more reliable analytics outputs that support better HR decision-making processes.

Proper data structuring is fundamental to AI model success in HR applications. Data normalization standardizes HR information to common formats, significantly improving model accuracy and consistency across different data sources. Feature engineering involves extracting and constructing relevant variables from raw HR data, creating meaningful inputs that enhance model insights and predictive capabilities. Transforming unstructured data, such as text documents, emails, and images, into structured formats enables AI processing and analysis. These techniques collectively prepare HR data for optimal AI consumption, ensuring models can effectively learn patterns and generate valuable insights.

Automation plays a crucial role in modern data preparation workflows, offering significant advantages over manual processes. Automated data cleaning tools expedite the process substantially, saving valuable time and reducing manual effort requirements across HR teams. By minimizing human intervention, automation reduces errors caused by manual data handling, ensuring higher data quality and reliability in AI-ready datasets. Additionally, automation helps maintain continuous data pipelines that support ongoing AI model training and deployment. This continuous processing capability is essential for maintaining model performance and ensuring AI systems remain current with evolving HR data patterns.

Understanding bias sources in HR datasets is critical for developing fair and effective AI systems. Historical hiring practices significantly influence current datasets, often embedding existing organizational biases into AI training data and perpetuating discriminatory patterns. Underrepresentation of certain demographic groups skews dataset fairness and accuracy, leading to AI models that may not perform equitably across all populations. Subjective evaluations introduce personal opinions and unconscious biases, affecting data objectivity and subsequent AI outcomes. Additionally, inconsistent or flawed data collection methods contribute to biased datasets, making systematic evaluation of data sources essential for responsible AI implementation.

Several proven methods exist for detecting and mitigating bias in HR AI systems. Statistical audits systematically analyze data patterns to identify bias within models and datasets, providing quantifiable evidence of fairness issues. Fairness metrics offer standardized ways to measure equality in AI decision-making processes, enabling organizations to track progress toward equitable outcomes. Data balancing techniques adjust datasets to ensure representative samples across demographic groups, reducing bias through improved data distribution. Continuous algorithm adjustments and ongoing monitoring systems help identify and address emerging bias over time, ensuring sustained fairness in AI applications throughout their operational lifecycle.

Biased data can have severe consequences for AI outcomes in HR applications, potentially damaging both individual careers and organizational reputation. Discriminatory hiring practices may result when AI systems trained on biased data unfairly exclude qualified candidates, perpetuating historical discrimination and limiting organizational diversity. Unfair performance evaluations represent another significant risk, as AI systems may produce skewed reviews that negatively impact employee growth, morale, and career advancement opportunities. Organizations implementing biased AI systems risk losing employee trust and damaging their reputation if decisions are perceived as unfair, making proactive bias mitigation essential for maintaining organizational integrity.

Data privacy regulations significantly impact how organizations handle HR data for AI applications. GDPR and CCPA represent core regulatory frameworks that govern data privacy and impose strict rules on HR data collection, processing, and storage practices. Transparent data collection methods and obtaining clear employee consent are fundamental requirements for compliance with privacy laws. Organizations must implement proper secure storage protocols and ensure lawful processing of HR data to protect employee rights and avoid regulatory penalties. Understanding these regulatory requirements is essential before implementing AI systems that process employee personal information.

GDPR compliance requires implementing several key best practices for HR data handling. Data minimization principles dictate collecting and processing only data necessary for specific, legitimate purposes, reducing privacy risks and enhancing employee protection. Secure data storage involves implementing robust security measures to protect personal information from unauthorized access, breaches, and other security threats. Explicit consent requirements mandate obtaining clear, informed permission from employees before collecting or processing their personal data for AI applications. Additionally, organizations must enable data access rights, allowing employees to view, correct, or delete their personal information in compliance with GDPR regulations.

Successfully balancing data utility with privacy and ethical considerations requires careful strategic planning and implementation. Organizations leverage HR data to gain valuable AI insights that improve workforce management and decision-making capabilities, driving business value through data-driven approaches. However, respecting employee privacy remains paramount, as maintaining appropriate safeguards preserves employee trust and ensures compliance with ethical and legal standards. Ethical data usage promotes sustainable AI deployment while strengthening organizational reputation and maintaining stakeholder trust. This balance requires ongoing attention to both technological capabilities and human values, ensuring AI implementations serve both organizational objectives and employee welfare.

Successful data integration requires connecting diverse data sources to create comprehensive insights that drive effective decision-making. Ensuring data quality and mitigating bias are critical for achieving fair and accurate AI outcomes that serve all employees equitably. Maintaining strict privacy compliance safeguards employee data throughout AI implementation processes, protecting individual rights while enabling organizational innovation. These foundational elements collectively enable responsible AI deployment that enhances HR decision-making processes while maintaining ethical standards and regulatory compliance. Thank you for participating in this training session.

Chapter 3: HR AI Tool Landscape

AI adoption in HR is revolutionizing how organizations manage their workforce through three primary benefits. First, automation of repetitive tasks allows AI to handle routine administrative work, significantly increasing operational efficiency and freeing HR professionals to focus on strategic, value-added activities like employee development and organizational planning. Second, enhanced recruitment accuracy leverages AI's ability to analyze vast amounts of candidate data, identifying optimal job-role matches more precisely than traditional methods. Finally, improved employee experience emerges through AI-driven systems that provide personalized support, responsive engagement tools, and tailored workplace solutions that boost overall employee satisfaction and retention rates.

While AI tools offer substantial benefits, they also present important challenges that organizations must address. Improved efficiency remains a primary advantage, as AI streamlines HR processes, reduces manual workloads, and accelerates decision-making timelines. Unbiased decision-making represents another key benefit, with AI systems analyzing data objectively without human prejudices, promoting fairness in hiring and evaluation processes. However, data privacy concerns require careful attention, as HR AI tools must implement robust security measures to protect sensitive employee information and comply with regulations. Additionally, ethical AI use demands transparency, accountability, and fairness in all algorithmic decisions to maintain trust and prevent discriminatory outcomes.

HR AI solutions fall into four key categories, each addressing specific organizational needs. Talent intelligence platforms leverage artificial intelligence to identify, assess, and strategically manage talent pools, providing comprehensive workforce insights for informed decision-making. Generative AI assistants automate routine HR tasks including responding to employee queries, scheduling interviews, and drafting communications, significantly reducing administrative burden. Resume parsing and matching engines utilize AI algorithms to analyze candidate resumes, extract relevant skills and experiences, and match applicants to job requirements with remarkable efficiency. People analytics platforms harness data analytics capabilities to enhance workforce planning, predict employee behavior patterns, and develop targeted engagement strategies that improve overall organizational performance.

Talent intelligence platforms offer comprehensive capabilities that transform strategic HR decision-making. Talent market mapping provides organizations with detailed insights into available skills distributions, workforce trends, and competitive talent landscapes across different geographic regions and industries. Skills gap analysis systematically identifies discrepancies between current employee capabilities and future organizational needs, enabling proactive upskilling initiatives and targeted hiring strategies. Competitor benchmarking delivers strategic intelligence by comparing talent acquisition metrics, compensation packages, and workforce composition against industry peers. Predictive hiring analytics utilize historical data and machine learning algorithms to forecast recruitment needs, predict candidate success rates, and optimize talent acquisition efforts for maximum organizational impact.

Talent intelligence platforms significantly enhance recruitment and retention strategies through data-driven insights. These systems provide comprehensive insight into candidate pools by analyzing demographic trends, skill distributions, geographic concentrations, and availability patterns, enabling more targeted and effective recruitment campaigns. Workforce trend analysis examines industry patterns, emerging skill demands, and labor market dynamics to help organizations align their talent strategies with future market conditions. Retention strategy design leverages predictive analytics to identify at-risk employees, understand departure triggers, and develop personalized interventions that reduce turnover rates. This data-driven approach enables HR teams to make proactive decisions that strengthen both talent acquisition effectiveness and employee retention outcomes.

Chapter 3: HR AI Tool Landscape

Leading talent intelligence solutions demonstrate the power of AI-driven workforce analytics through several key features. AI-driven talent analytics provide deep insights into talent pools by analyzing skill trends, geographic distributions, career progression patterns, and market dynamics, enabling organizations to make informed strategic workforce decisions. Comprehensive sourcing capabilities offer advanced tools that streamline candidate identification across multiple platforms, social networks, and professional databases, significantly expanding recruitment reach and effectiveness. Notable platforms in this space combine sophisticated analytics engines with powerful sourcing functionalities, creating integrated solutions that optimize both hiring processes and long-term workforce planning strategies. These platforms represent the cutting edge of talent intelligence technology in modern HR practice.

Generative AI applications in HR span multiple critical functions, revolutionizing workplace communication and administrative efficiency. Email drafting capabilities enable AI to automatically generate professional correspondence, including recruitment communications, policy announcements, and employee notifications, maintaining consistent tone and messaging while saving significant time. Employee query response systems provide instant, accurate answers to common HR questions about benefits, policies, and procedures, improving employee satisfaction and reducing HR workload. Training content generation utilizes AI to create customized learning materials, interactive modules, and assessment tools tailored to specific roles and skill levels. Performance review assistance helps managers document evaluations more effectively, suggesting relevant feedback language and ensuring comprehensive, constructive assessments.

Automating employee communication and support through AI transforms HR service delivery and accessibility. 24/7 employee support ensures round-the-clock assistance through intelligent chatbots that handle routine inquiries, process basic requests, and escalate complex issues appropriately, eliminating time zone limitations and response delays. Streamlined onboarding leverages virtual assistants to guide new hires through documentation, training schedules, policy reviews, and system setups, creating consistent and efficient integration experiences. Automated FAQ systems reduce HR intervention by providing instant responses to frequently asked questions about benefits, policies, procedures, and workplace guidelines. These automation capabilities collectively reduce HR workload significantly, allowing human resources professionals to focus on strategic initiatives, complex problem-solving, and meaningful employee engagement activities.

Generative AI transforms onboarding, training, and performance management through personalized, efficient solutions. Personalized onboarding plans utilize AI to customize integration experiences based on individual employee roles, experience levels, learning preferences, and departmental requirements, ensuring relevant and engaging first-day experiences. Custom training modules leverage AI generation capabilities to create tailored educational content that adapts to different skill levels, learning speeds, and professional development goals, maximizing training effectiveness and employee engagement. Performance review summaries assist managers by analyzing performance data, identifying key achievements and improvement areas, and generating comprehensive, balanced evaluation narratives that facilitate meaningful conversations between supervisors and employees, ultimately improving the quality and consistency of performance management processes.

Automated resume screening and ranking revolutionize recruitment efficiency through intelligent data processing. AI resume parsing technology efficiently extracts relevant information from candidate applications, including skills, experiences, education credentials, and employment history, organizing this data into standardized formats for easy comparison and analysis. Candidate ranking systems evaluate applicants based on predetermined criteria such as skill alignment, experience relevance, and qualification matches, automatically prioritizing the most suitable candidates for specific positions. Bias reduction represents a critical advantage, as automated screening processes eliminate human prejudices related to names, backgrounds, or demographic characteristics, promoting fair and objective candidate evaluation. This systematic approach ensures consistent evaluation standards while significantly reducing time-to-hire and improving recruitment outcomes.

Matching algorithms for job-candidate fit utilize sophisticated analysis techniques to optimize hiring decisions. Skill and experience analysis evaluates candidates' technical competencies, professional backgrounds, and achievement records against specific job requirements, creating detailed compatibility scores that highlight the strongest matches. These algorithms examine years of experience, skill proficiency levels, industry knowledge, and previous role responsibilities to determine objective fitness ratings. Cultural fit assessment goes beyond technical qualifications to evaluate compatibility between candidate values, work styles, communication preferences, and organizational culture. This comprehensive approach considers personality traits, team dynamics, leadership styles, and company values alignment to predict successful long-term integration and employee satisfaction within the existing workplace environment.

Integration with applicant tracking systems creates seamless, end-to-end recruitment automation. Seamless ATS integration enables resume parsing engines to connect directly with existing recruitment platforms, automatically populating candidate profiles, maintaining data consistency, and eliminating manual data entry requirements across systems. End-to-end automation encompasses the entire recruitment workflow from initial application receipt through final candidate selection, including automated screening, ranking, communication scheduling, and status updates. Enhanced recruitment efficiency results from this integrated approach, as automated parsing and matching capabilities significantly improve the speed and accuracy of candidate screening processes within ATS environments. This comprehensive integration reduces administrative overhead, minimizes human error, and accelerates hiring timelines while maintaining high-quality candidate evaluation standards.

People analytics platforms provide data-driven insights essential for strategic workforce planning and optimization. Workforce demographics analysis examines employee composition across various dimensions including age, gender, tenure, education levels, and geographic distribution, revealing diversity trends and helping organizations understand their talent landscape comprehensively. Skills and performance evaluation utilizes analytical tools to assess employee capabilities, identify high performers, track development progress, and pinpoint areas requiring improvement or additional training investment. Optimizing hiring and succession planning leverages predictive analytics to forecast future talent needs, identify potential leadership candidates, and develop strategic succession pathways. These data-driven insights enable organizations to make informed decisions about talent acquisition, development investments, and organizational restructuring initiatives.

Predictive analytics for employee engagement and retention enables proactive workforce management strategies. Identifying at-risk employees utilizes machine learning algorithms to analyze various data points including performance metrics, engagement survey results, attendance patterns, and behavioral indicators to predict which employees may be considering departure. These models examine factors such as job satisfaction scores, career progression rates, compensation equity, and workplace relationships to generate risk assessments. Targeted retention strategies emerge from these predictive insights, enabling HR teams to develop personalized interventions including career development opportunities, compensation adjustments, role modifications, or enhanced support systems. This proactive approach allows organizations to address potential retention issues before employees actually leave, significantly reducing turnover costs and maintaining organizational knowledge.

Chapter 3: HR AI Tool Landscape

Ethical considerations and data privacy in people analytics require careful attention to maintain employee trust and legal compliance. Ethical standards importance emphasizes the need for transparent, fair, and accountable use of employee data in analytics processes, ensuring that algorithmic decisions respect individual rights and promote equitable treatment across all employee groups. Data privacy regulations mandate strict compliance with laws such as GDPR, CCPA, and other regional privacy requirements, necessitating robust security measures, consent management, and data minimization practices in analytics implementations. Maintaining employee trust depends on transparent communication about data usage, clear opt-out mechanisms, and demonstrable benefits from analytics initiatives. Organizations must balance analytical insights with privacy protection, ensuring that people analytics enhance rather than undermine workplace relationships.

In conclusion, HR AI tools are fundamentally transforming human resources through four key areas of impact. Automation of HR processes streamlines repetitive administrative tasks, reduces manual workloads, and increases operational efficiency, allowing HR professionals to focus on strategic initiatives and employee development. Enhanced decision-making emerges from data-driven AI insights that support better hiring decisions, performance evaluations, and strategic workforce planning based on objective analytics rather than intuition alone. Improved employee experience results from AI-powered personalization of HR services, responsive support systems, and tailored engagement strategies that boost workplace satisfaction. Building agile and innovative HR functions requires embracing AI technologies that enable organizations to respond quickly to changing workforce needs, adapt to market conditions, and maintain competitive advantages in talent management.

Chapter 4: Intelligent Candidate Sourcing

Intelligent candidate sourcing brings significant benefits but also presents important challenges we must address. The primary advantage is increased efficiency - AI streamlines recruitment processes, saving valuable time and resources while simultaneously improving the quality of candidates identified. However, we face data privacy concerns as handling candidate information raises critical security and compliance issues. Additionally, technology integration challenges emerge when implementing intelligent sourcing tools into existing recruitment workflows, requiring careful planning and seamless adaptation to ensure successful adoption.

Modern AI technologies revolutionize sourcing strategies through three key capabilities. Automated candidate identification enables AI to quickly and efficiently process large datasets, identifying potential candidates at scale. Machine learning algorithms enhance matching accuracy by analyzing skills, experience, and job requirements with greater precision than traditional methods. Natural language processing supports enhanced decision-making by analyzing candidate data and communication patterns, providing recruiters with deeper insights into candidate suitability and enabling more informed hiring decisions throughout the recruitment process.

AI-powered personalization transforms how we create and deliver job advertisements to potential candidates. Tailored job ads leverage artificial intelligence to customize content based on individual candidate preferences and behaviors, resulting in significantly better engagement rates. This targeted messaging approach increases relevance by ensuring job advertisements reach suitable candidates more effectively. The result is improved response rates as AI-driven personalization delivers meaningful and engaging job advertisements that resonate with candidates' interests, career goals, and professional backgrounds, ultimately attracting higher-quality applicants.

Optimizing advertisement placements requires strategic data analysis and audience targeting for maximum impact. AI analyzes comprehensive data patterns to identify optimal timing and platforms for ad placements, ensuring maximum visibility when candidates are most active. Target audience segmentation allows advertisements to reach the most suitable candidate segments, improving overall recruitment outcomes and return on investment. This systematic approach to maximizing ad visibility ensures higher engagement rates from potential candidates by placing the right message in front of the right audience at the optimal time.

Measuring and analyzing engagement effectiveness provides crucial insights for continuous improvement in recruitment advertising. AI-driven insights capture detailed engagement metrics, enabling recruiters to make data-informed decisions about campaign performance and optimization opportunities. Campaign refinement processes use these insights to optimize targeting strategies, reaching the right audience more effectively and improving overall campaign effectiveness. This analytical approach leads to improved return on investment as tracking and refining recruitment advertisements results in better resource allocation and more successful talent acquisition outcomes.

Identifying passive candidates requires sophisticated data analysis techniques to uncover hidden talent pools. Behavioral data analysis enables AI to process candidate information and identify potential candidates who aren't actively seeking new opportunities but may be open to the right position. Online footprint evaluation provides valuable insights into candidate skills, interests, and professional activities through their digital presence. This approach significantly enhances recruitment reach by helping recruiters connect with qualified passive candidates who traditional recruitment methods might overlook, expanding the available talent pool considerably.

Automated outreach strategies streamline communication with passive candidates while maintaining personalization and timing effectiveness. AI-powered messaging tools automate outreach by generating and sending personalized messages to passive candidates, ensuring each communication feels relevant and engaging rather than generic or mass-produced. Timely communication ensures that automated outreach happens at optimal moments when candidates are most likely to be receptive and respond positively. This systematic approach increases response rates while reducing manual effort required from recruitment teams.

Enhancing candidate engagement requires sophisticated AI-driven insights to understand and respond to individual candidate preferences. Predicting candidate interests involves analyzing available data to forecast what motivates and interests potential candidates, enabling more personalized and effective engagement strategies. Optimizing communication preferences helps identify the most effective channels and communication styles for different candidates, significantly improving response rates from passive candidates. Using these AI insights enables nurturing strong, lasting relationships with passive candidates, creating a valuable talent pipeline for future recruitment success.

AI-based assessment provides comprehensive evaluation of both technical and interpersonal skills for holistic candidate evaluation. Technical competency evaluation uses data-driven analysis and simulations to assess candidates' hard skills accurately and objectively, ensuring technical requirements are met. Interpersonal skills assessment evaluates soft skills through behavioral simulations and communication pattern analysis, providing insights into collaboration, leadership, and cultural fit potential. This comprehensive approach creates complete candidate profiles by combining hard and soft skills data, enabling better-informed hiring decisions that consider all aspects of candidate suitability.

Predicting organizational culture compatibility helps ensure successful long-term hires and positive workplace integration. AI-driven compatibility analysis evaluates candidate traits, values, and working styles against established company culture metrics to predict integration success and overall fit. This predictive approach helps reduce employee turnover by selecting candidates who naturally align with organizational values and working environments. Better cultural fit promotes improved team dynamics and workplace harmony, creating more cohesive teams and positive working relationships that benefit both individual employees and organizational performance overall.

Predictive analytics transforms hiring decisions by providing data-driven insights into candidate potential and long-term success probability. AI-driven performance forecasting uses advanced algorithms to predict candidate job performance before making hiring decisions, reducing uncertainty in the selection process. Retention likelihood prediction helps identify candidates most likely to remain with the organization long-term, reducing costly turnover and improving hiring quality. This data-driven approach enables recruiters to make confident hiring choices based on comprehensive insights from predictive analytics tools, improving overall recruitment success rates.

AI-powered candidate sourcing represents the future of recruitment, offering significant advantages for organizations seeking top talent. This technology improves recruitment efficiency by sourcing qualified candidates from large talent pools quickly and accurately. Enhanced recruitment accuracy ensures better matching between candidates and job requirements, reducing hiring mistakes and improving employee satisfaction. AI-powered tools enable more effective candidate engagement through personalized communication strategies. Adopting intelligent sourcing technologies is becoming crucial for competitive talent acquisition, helping organizations attract and retain top talent in an increasingly competitive market environment.

Chapter 5: Resume Screening and Shortlisting

Traditional resume screening relies heavily on manual processes where recruiters examine hundreds of applications, focusing primarily on keyword matching, qualifications, and experience levels. This approach, while familiar, presents significant challenges in today's competitive hiring landscape. Manual review is inherently time-consuming, often taking minutes per resume when dealing with large applicant volumes. The process also suffers from consistency issues, as subjective interpretations can vary between reviewers, leading to inconsistent candidate evaluations. These manual methods, though thorough, struggle to keep pace with modern recruitment demands and can inadvertently introduce bias into the selection process.

Manual shortlisting faces three critical challenges that impact hiring quality and efficiency. Human error becomes inevitable when recruiters process large volumes of applications, leading to fatigue-induced mistakes and oversight of qualified candidates. Unconscious bias represents a significant concern, as personal preferences and preconceptions can influence candidate selection, resulting in unfair hiring practices that limit diversity. Scalability issues emerge when organizations need to handle hundreds or thousands of applications manually, creating bottlenecks that delay hiring decisions and strain recruitment resources. These limitations highlight the urgent need for automated, objective screening solutions in modern recruitment.

Natural Language Processing revolutionizes recruitment by transforming unstructured resume text into actionable, structured data that streamlines candidate screening processes. NLP techniques automatically extract key information such as skills, experience, education, and qualifications from diverse resume formats, eliminating the need for manual data entry. This automated approach significantly improves recruitment efficiency by rapidly identifying relevant applicant information and enabling instant comparisons across large candidate pools. By leveraging NLP, organizations can process hundreds of resumes in minutes rather than hours, while maintaining consistency and accuracy in data extraction and candidate evaluation.

Modern NLP systems employ three sophisticated techniques to extract structured data from resumes effectively. Named Entity Recognition identifies and categorizes key information such as candidate names, employment dates, company names, educational institutions, and geographic locations from unstructured text. Resume content parsing breaks down complex documents into meaningful components, organizing information into distinct sections like work experience, education, skills, and achievements for easier analysis. Text classification assigns relevant categories to resume content, enabling automated sorting and filtering based on job requirements, industry sectors, or specific skill sets, facilitating more efficient candidate matching processes.

Automated resume parsing offers significant advantages while presenting certain challenges that organizations must address. The primary benefits include dramatically enhanced speed and consistency in processing applications, with systems capable of parsing hundreds of resumes per hour while maintaining uniform evaluation criteria. However, challenges arise with format variations, as diverse resume layouts, fonts, and structures can affect parsing accuracy. Language nuances and contextual understanding remain limitations, as automated systems may struggle with industry-specific terminology, creative job descriptions, or non-standard formatting. Continuous improvement through machine learning updates and regular system refinements is essential for maintaining optimal parsing performance.

Candidate-job matching algorithms operate on three fundamental principles that enable precise alignment between applicant qualifications and position requirements. Keyword matching forms the foundation, comparing specific terms and phrases between resumes and job descriptions to identify relevant skills, technologies, and experience areas. Semantic analysis advances beyond simple keyword detection by examining the contextual meaning and relationships between words, understanding synonyms, related concepts, and industry terminology. Machine learning models continuously evolve by learning from successful hiring decisions and recruiter feedback, improving prediction accuracy over time and adapting to changing job market trends and organizational preferences.

Effective ranking systems transform raw matching data into actionable insights for recruiters through systematic candidate evaluation. Candidate scoring assigns numerical values based on relevance, qualifications, experience level, and alignment with job requirements, creating objective assessment metrics. The ranking process prioritizes candidates with the highest compatibility scores, ensuring that recruiters focus their attention on the most promising applicants first. This systematic approach significantly enhances recruiter efficiency by eliminating time spent reviewing unsuitable candidates and enabling rapid identification of top talent. Automated ranking also provides consistency across different recruiters and hiring managers, reducing subjective variations in candidate evaluation.

Integration with Applicant Tracking Systems creates seamless workflow automation that transforms the entire recruitment process. Modern ATS platforms incorporate sophisticated matching algorithms directly into existing recruitment workflows, eliminating the need for separate tools or manual data transfers. This integration significantly enhances recruiter productivity by automating routine screening tasks, generating candidate rankings, and providing intelligent recommendations for interview selection. The streamlined process also improves candidate experience through faster response times, more accurate job matching, and reduced application processing delays. Automated workflows ensure that no qualified candidates are overlooked while maintaining comprehensive audit trails for compliance purposes.

Recruitment bias manifests in multiple forms that can significantly impact fair hiring practices and workplace diversity. Gender bias often influences evaluation of identical qualifications differently based on perceived gender associations with certain roles or industries. Ethnicity bias can affect candidate assessment through name recognition, cultural assumptions, or unconscious preferences for familiar backgrounds. Age bias may lead to discrimination against both younger candidates perceived as inexperienced and older candidates assumed to be less adaptable to new technologies. Educational background bias can unfairly favor prestigious institutions over practical experience or alternative educational paths, potentially overlooking highly qualified candidates from diverse learning environments.

Algorithmic approaches offer powerful tools for minimizing bias and promoting equitable recruitment practices. Resume anonymization removes personal identifiers such as names, photos, addresses, and demographic indicators, enabling evaluation based solely on qualifications and experience. Fairness-aware machine learning models incorporate specific constraints designed to prevent discriminatory outcomes by monitoring for disparate impact across protected groups. Continuous bias monitoring systems track hiring patterns, identify potential discrimination, and alert administrators to concerning trends in real-time. These technological solutions work together to create objective, data-driven recruitment processes that promote equality while maintaining high standards for candidate selection.

Implementing fair and equitable candidate evaluation requires a comprehensive approach combining technology with human judgment. Human oversight ensures that empathy, cultural understanding, and contextual assessment remain integral to hiring decisions, preventing over-reliance on automated systems. Unbiased algorithms, designed with fairness principles and regularly audited for discriminatory patterns, provide consistent, objective candidate assessment across all applications. Standardized evaluation criteria establish clear, measurable metrics that apply equally to all candidates, promoting transparency and accountability. Open communication about evaluation methods builds trust with candidates and stakeholders while demonstrating organizational commitment to fair hiring practices and regulatory compliance.

Natural Language Processing revolutionizes resume screening by automatically extracting and analyzing candidate information with unprecedented speed and accuracy. Sophisticated matching algorithms enhance recruitment precision by intelligently aligning candidate qualifications with job requirements, while ranking systems prioritize the most suitable applicants. Most importantly, bias reduction techniques promote fairness and diversity by minimizing unconscious discrimination and creating more equitable hiring processes. By implementing these advanced techniques, organizations can achieve faster, more accurate, and fairer recruitment outcomes that benefit both employers and candidates in today's competitive talent market.

Chapter 6: AI-Assisted Interviewing

This section introduces the transformative power of AI in recruitment processes. We'll explore how artificial intelligence is fundamentally changing the way organizations approach talent acquisition. The integration of AI technologies addresses traditional recruitment challenges including time constraints, unconscious bias, and inconsistent evaluation methods. These advancements represent a significant shift toward data-driven decision making in human resources. Understanding these changes is crucial for modern HR professionals and organizational leaders.

AI-assisted interviewing brings significant benefits alongside important challenges that organizations must address. Enhanced objectivity reduces human biases while standardizing candidate evaluation processes. Faster recruitment cycles enable organizations to secure top talent before competitors. However, data privacy concerns require careful consideration of how candidate information is collected, stored, and used. Algorithmic bias and transparency issues demand ongoing monitoring and adjustment of AI systems. Organizations must balance technological efficiency with ethical responsibility and regulatory compliance to maximize AI's benefits while minimizing potential risks.

Video interview analysis and scoring represents one of the most sophisticated applications of AI in recruitment. This technology enhances objectivity by providing standardized evaluation criteria and reducing subjective interpretation. Advanced algorithms analyze multiple data points simultaneously, including verbal responses, facial expressions, and body language. This comprehensive approach offers deeper insights into candidate suitability beyond traditional interview methods. The technology enables more accurate predictions of job performance and cultural fit while maintaining consistency across all evaluations.

Automated video assessment technologies leverage two primary AI capabilities to evaluate candidates comprehensively. Computer vision analysis evaluates facial expressions and body language in real-time, providing insights into candidate confidence, engagement, and authenticity. Natural language processing assesses candidate responses by analyzing speech content, tone, and linguistic patterns. These technologies work together to create a complete picture of candidate performance. The real-time nature of this analysis allows for immediate feedback and scoring, significantly reducing evaluation time while increasing accuracy and consistency.

Scoring algorithms utilize sophisticated criteria to evaluate candidates objectively and consistently. Evaluation of verbal cues analyzes spoken responses for clarity, relevance, and content quality, ensuring candidates can communicate effectively. Assessment of non-verbal cues examines body language and facial expressions to gauge confidence and engagement levels. Standardized candidate scoring uses predefined criteria to ensure fair and consistent assessments across different interviewers and sessions. This systematic approach eliminates variability in evaluation standards and provides comparable scores that support data-driven hiring decisions.

Reducing bias in candidate evaluation requires systematic approaches and continuous improvement processes. Minimizing unconscious bias involves AI systems focusing on objective data rather than subjective impressions that may be influenced by personal preferences. Continuous monitoring ensures fairness by tracking system performance and identifying potential biases over time. Tuning for fairness requires regular updates and adjustments to AI algorithms to maintain accuracy and improve evaluation quality. This ongoing process helps organizations maintain ethical hiring practices while leveraging AI's analytical capabilities for better decision-making.

Behavioral and sentiment analysis represents the next frontier in understanding candidate responses beyond surface-level answers. This sophisticated technology examines micro-expressions, speech patterns, and emotional indicators to provide deeper insights into candidate personality and fit. By analyzing both conscious and unconscious behavioral cues, organizations gain a more complete understanding of potential employees. This approach helps identify candidates who not only have the right skills but also demonstrate the emotional intelligence and behavioral traits necessary for success in specific roles.

Advanced techniques for behavioral analysis provide unprecedented insights into candidate psychology and communication patterns. Micro-expression analysis detects subtle facial expressions that reveal true emotions beyond spoken words, helping identify authenticity and emotional responses. Speech pattern recognition evaluates tone, pace, and hesitations to assess confidence and honesty levels during responses. Gesture interpretation analyzes body language and hand movements to measure engagement and sincerity. These combined techniques create a comprehensive behavioral profile that enhances traditional interview evaluation methods and provides deeper candidate insights.

Sentiment detection and interpretation technologies provide sophisticated analysis of candidate emotional states and intentions. Sentiment analysis algorithms process verbal cues to assess emotional tone accurately and efficiently, identifying enthusiasm, confidence, or concern in responses. Non-verbal cues evaluation enhances understanding by analyzing facial expressions and gestures that may contradict or confirm spoken words. Understanding true intentions helps reveal candidates' genuine feelings and motivations beyond prepared responses. This comprehensive emotional analysis supports more informed hiring decisions by identifying candidates whose values and attitudes align with organizational culture.

Applications in candidate profiling demonstrate how behavioral and sentiment data enhance recruitment decision-making. Behavioral data insights reveal candidate actions, preferences, and work habits that traditional interviews might miss, creating more accurate profiles. Sentiment analysis plays a crucial role in gauging candidate attitudes and emotional responses, enabling interviewers to tailor their approach accordingly. Improved hiring decisions result from combining multiple data types to enable informed choices and personalized interview strategies. This comprehensive profiling approach leads to better job matches and reduced employee turnover rates.

Real-time coaching for interviewers represents a significant advancement in improving interview quality and consistency. This technology provides immediate feedback and guidance during interviews, helping interviewers optimize their approach dynamically. AI-powered coaching systems analyze interview flow, question effectiveness, and candidate responses to suggest improvements in real-time. This support enhances interviewer performance while maintaining the human element essential for building rapport with candidates. The result is higher-quality interviews that better assess candidate suitability while providing positive candidate experiences.

AI-powered feedback mechanisms provide comprehensive support to enhance interviewer performance during candidate evaluations. Real-time suggestions offer instant feedback on question phrasing to improve clarity and engagement, ensuring optimal communication with candidates. Adjusting interaction pace helps interviewers monitor and modify conversation flow for better communication dynamics and candidate comfort. Enhancing interaction style provides guidance on interviewing techniques, promoting effective and adaptive approaches that respond to individual candidate needs. These mechanisms ensure consistent high-quality interviews while supporting interviewer development and confidence.

Best practices for interviewer support emphasize the balance between technological assistance and human judgment in recruitment processes. Integrating AI insights involves using data-driven information to assist interviewers in making informed decisions while maintaining personal connection. Human judgment importance cannot be overstated - maintaining empathy and intuition ensures fair and balanced assessments that consider factors beyond data points. Error reduction leverages technology to minimize biases and mistakes during evaluations, resulting in better hiring outcomes. Successful implementation requires training interviewers to effectively utilize AI tools while preserving human-centered evaluation approaches.

The impact on decision-making and candidate experience demonstrates AI's transformative effect on recruitment outcomes. Improved decision accuracy results from AI support that enhances precision by analyzing data objectively, reducing errors in candidate selection. Fair interview processes emerge from consistent AI-driven performance standards that ensure fairness and reduce bias across all evaluations. Engaging candidate experience is created through AI's ability to provide more transparent, efficient, and responsive interview processes that improve candidate satisfaction. These improvements contribute to enhanced employer branding and better recruitment outcomes for organizations.

In conclusion, AI-assisted interviewing transforms recruitment through four key advancements. Enhanced objectivity removes bias and promotes fair evaluation, ensuring all candidates receive equitable treatment. Deeper candidate insights provide comprehensive analysis of skills and potential beyond traditional assessment methods. Real-time interview support offers AI-driven suggestions and feedback during interviews, improving quality and consistency. Future hiring efficiency promises continued evolution of AI technologies toward fairer and more efficient recruitment processes. Organizations adopting these technologies gain competitive advantages in talent acquisition while building more effective, unbiased hiring practices.

Chapter 7: Recruitment Workflow Automation

Recruitment workflow automation represents a fundamental shift in how organizations approach hiring. By automating repetitive tasks like resume screening and initial candidate communications, technology reduces manual workload while minimizing human errors. This automation allows HR teams to focus their expertise on strategic activities like cultural fit assessment and final hiring decisions. The result is faster candidate processing through streamlined workflows and improved hiring consistency across all recruitment stages. Organizations implementing these systems see dramatic improvements in both efficiency and candidate satisfaction, creating a competitive advantage in talent acquisition markets.

The benefits of recruitment workflow automation extend far beyond simple task reduction. Organizations Wons experience significantly increased operational efficiency as automation handles routine processes around the clock. Hiring timelines compress dramatically through automated candidate screening and streamlined communication workflows. Candidates enjoy enhanced experiences with timely responses and simplified application processes that reflect modern digital expectations. Perhaps most importantly, automated systems ensure better compliance with employment regulations while improving data accuracy by reducing manual data entry errors. These combined benefits create measurable ROI for organizations investing in recruitment technology.

Chatbots have revolutionized candidate engagement by providing instant responses to frequently asked questions at any hour. This immediate availability enhances the candidate experience while significantly reducing the burden on recruiters who previously handled routine inquiries manually. Automated responses ensure candidates receive consistent, up-to-date information without delays that could lead to candidate drop-off. The technology maintains accuracy across all interactions while freeing recruiters to focus on complex candidate evaluations and relationship building. This efficiency improvement translates directly into cost savings and improved candidate satisfaction scores.

Modern recruitment chatbots go beyond simple FAQ responses to deliver truly personalized communication experiences. By analyzing individual candidate profiles and tracking their progress through the recruitment pipeline, chatbots customize messaging for maximum relevance and engagement. This tailored approach provides candidates with specific guidance related to their application status and next steps. Enhanced candidate support through personalized interactions demonstrates organizational professionalism and attention to detail. The result is improved candidate experience that strengthens employer branding and increases the likelihood of top candidates accepting job offers.

Effective chatbot implementation directly impacts candidate satisfaction and retention throughout the recruitment process. Consistent communication ensures candidates feel informed and valued, reducing anxiety about application status and next steps.

Chapter 7: Recruitment Workflow Automation

Responsive chatbot interactions maintain candidate engagement during potentially lengthy recruitment processes, preventing top talent from pursuing other opportunities. This improved communication strategy significantly reduces candidate drop-off rates and boosts retention in the recruitment pipeline. Organizations using chatbots report higher candidate satisfaction scores and improved completion rates through all recruitment stages.

Automated interview scheduling eliminates the traditional back-and-forth communication that often creates delays and frustration. Candidates can independently select interview times based on recruiter availability, providing flexibility and convenience that modern job seekers expect. This self-service approach reduces administrative burden on HR teams while minimizing scheduling errors and conflicts. The system automatically handles calendar conflicts and sends confirmations, creating a professional experience that reflects well on the organization. Reduced scheduling friction means faster time-to-hire and improved candidate experience throughout the interview process.

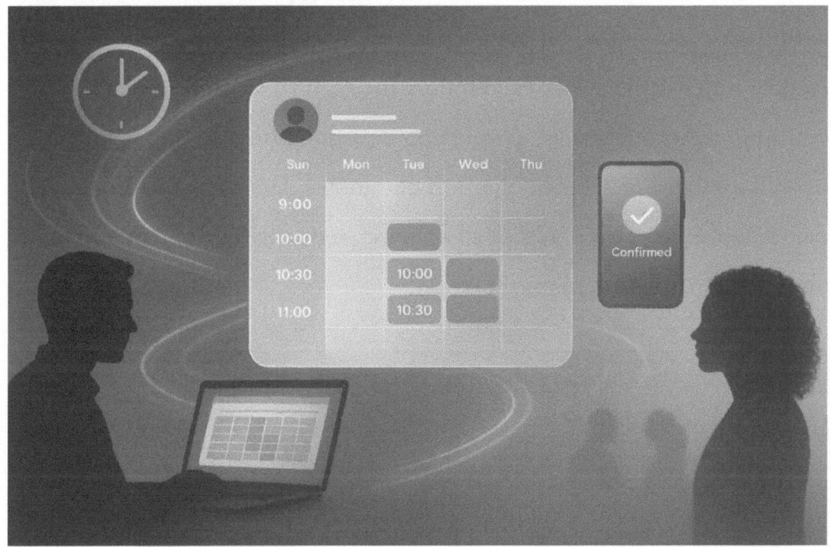

Real-time notifications ensure all parties stay informed throughout the interview scheduling process. Candidates and recruiters receive instant confirmations, reminders, and updates about interview arrangements, reducing the likelihood of missed appointments. This improved coordination helps synchronize busy schedules between multiple stakeholders, including hiring managers, panel interviewers, and candidates. Automated notifications can include interview details, location information, and preparation materials, ensuring everyone arrives prepared and on time. The result is smoother interview processes with fewer last-minute complications or rescheduling requirements.

Integration with existing calendar and communication tools creates seamless workflows that improve efficiency for all users. Scheduling updates automatically appear across all platforms, ensuring everyone has access to current information without manual data entry. Recruiters benefit from hassle-free scheduling processes that integrate with their existing productivity tools and workflows. Candidates receive timely updates and reminders through their preferred communication channels, simplifying coordination and improving overall experience. This integration eliminates data silos and ensures consistent information across all systems, reducing confusion and improving professional presentation.

Recruitment funnel analytics provide crucial visibility into candidate movement through each stage of the hiring process. HR teams can monitor candidate flow to identify where progression occurs smoothly and where bottlenecks develop. Conversion rate measurements highlight the percentage of candidates advancing versus those dropping off at each stage, revealing process effectiveness. This data-driven approach enables organizations to understand their recruitment performance objectively rather than relying on subjective impressions. Analytics reveal patterns in candidate behavior that can inform process improvements and resource allocation decisions.

Funnel data analysis enables organizations to detect specific delays that impact overall process efficiency and candidate experience. By pinpointing exact bottlenecks, HR teams can implement focused interventions to resolve issues causing hiring timeline delays. This targeted approach to process improvement is more effective than broad-based changes that may not address root causes. Data-driven insights guide strategic decisions about where to invest resources for maximum impact on recruitment outcomes. Organizations using this analytical approach see measurable improvements in time-to-hire and candidate satisfaction metrics.

Leveraging recruitment data transforms hiring from an intuitive process to a strategic, evidence-based function. Data-driven decisions rely on accurate analysis and measurable insights rather than subjective impressions or traditional practices. This information empowers strategic workforce planning and optimizes resource allocation based on demonstrated effectiveness rather than assumptions. Continuous use of recruitment data drives ongoing enhancements in hiring processes and outcomes through iterative improvement cycles. Organizations embracing data-driven recruitment see improved hiring quality, reduced time-to-fill positions, and better alignment between recruitment activities and business objectives.

Chapter 7: Recruitment Workflow Automation

Recruitment workflow automation represents a comprehensive solution that addresses multiple challenges facing modern HR departments. By boosting efficiency through reduced manual tasks and accelerated hiring processes, organizations achieve measurable productivity gains. Enhanced candidate experience through improved communication and transparency creates competitive advantages in talent acquisition. Most importantly, data-driven insights enable smarter hiring decisions based on evidence rather than intuition. Organizations implementing these integrated automation solutions report significant improvements in hiring quality, speed, and candidate satisfaction while reducing overall recruitment costs and administrative burden.

Chapter 8: Personalized Onboarding Experiences

Personalized onboarding represents a fundamental shift from one-size-fits-all approaches to tailored experiences that recognize each new hire's unique background, role requirements, and learning preferences. This section will establish the foundation for understanding why personalization matters and how it directly impacts both employee satisfaction and organizational success. We'll explore the core principles that drive effective personalization and examine the measurable benefits for both new hires and organizations.

Effective personalized onboarding rests on three fundamental principles that transform the new hire experience. Understanding employee needs begins with comprehensive assessment of each individual's background, expectations, and learning preferences. This assessment enables organizations to create targeted pathways that address specific knowledge gaps and career aspirations. Providing relevant content ensures that every piece of information shared directly applies to the employee's role and growth trajectory, eliminating irrelevant material that can overwhelm and confuse. Fostering engagement through interactive experiences creates meaningful connections between new hires and their teams, building confidence and accelerating productivity from the earliest stages of employment.

The benefits of personalized onboarding create a powerful value proposition for both organizations and new employees. Improved new hire retention occurs when employees feel valued and supported from their first day, reducing costly turnover that can reach up to 50% in the first year. Accelerated learning curves result from customized training programs that match individual learning styles and existing skill levels, enabling new hires to reach productivity benchmarks faster. Enhanced job satisfaction emerges when onboarding aligns roles with employee expectations and strengths, creating positive initial experiences that influence long-term engagement. Organizations benefit through increased productivity, decreased turnover costs, and stronger employer branding that attracts top talent.

Successfully implementing personalized onboarding requires careful consideration of several critical challenges. Data privacy issues demand robust security measures and transparent communication about how employee information is collected, stored, and used throughout the personalization process. Organizations must establish clear policies and obtain proper consent while maintaining trust. Resource allocation presents ongoing challenges as personalized approaches require greater investment in technology, training, and personnel compared to traditional methods. Balancing automation with human interaction ensures that efficiency gains don't come at the expense of meaningful personal connections that are essential for building company culture and employee relationships.

Artificial intelligence is revolutionizing onboarding by enabling unprecedented levels of personalization and adaptability. This section explores how AI technologies analyze employee data to create customized learning experiences that evolve in real-time based on individual progress and preferences. We'll examine the mechanisms through which AI personalizes content delivery, adjusts pacing, and enhances engagement throughout the onboarding journey,

Chapter 8: Personalized Onboarding Experiences

demonstrating how technology can create more human-centered experiences.

AI transforms onboarding through three key mechanisms that create truly personalized experiences. Data-driven personalization leverages comprehensive employee profiles, including background, role requirements, and learning preferences, to customize content and delivery methods for maximum relevance and effectiveness. Adaptive learning pacing uses machine learning algorithms to monitor individual progress and automatically adjust the speed and complexity of content delivery, ensuring optimal learning outcomes without overwhelming or under-challenging new hires. Enhanced engagement results from AI's ability to present information in formats that match individual preferences, whether visual, auditory, or interactive, while providing personalized recommendations and pathways that maintain motivation throughout the onboarding process.

Advanced AI capabilities enable sophisticated customization and real-time adaptation that continuously improves the onboarding experience. Real-time learning adaptation represents AI's ability to instantly analyze employee responses, quiz results, and engagement patterns to modify content difficulty, suggest additional resources, or redirect learning paths as needed. This dynamic adjustment ensures that each employee receives optimal support precisely when needed. Data-driven personalization goes beyond basic demographics to analyze behavioral patterns, learning preferences, and performance indicators, enabling AI systems to predict and proactively address potential challenges while identifying opportunities to accelerate learning and engagement through customized module sequencing and content presentation.

Real-world implementations of AI-driven onboarding demonstrate significant measurable improvements across key performance indicators. Improved onboarding efficiency manifests through reduced administrative burden on HR teams and shortened time-to-productivity for new hires, with some organizations reporting up to 40% reduction in onboarding completion time. Increased new hire satisfaction scores reflect the enhanced relevance and personalization of the experience, with many companies seeing satisfaction ratings improve by 25-30%. Higher retention rates directly correlate with effective AI-supported onboarding, as personalized experiences create stronger initial connections between employees and organizations, resulting in reduced first-year turnover and improved long-term employee engagement and performance.

Interactive learning bots represent a significant advancement in onboarding support, providing instant, personalized assistance that scales to accommodate any number of new hires simultaneously. This section examines how chatbots and virtual assistants enhance the onboarding experience by delivering immediate support, answering common questions, and guiding employees through complex processes with consistent, friendly interaction that's available 24/7.

Interactive learning bots serve three essential functions that transform onboarding support delivery. Instant support eliminates frustrating wait times by providing immediate responses to new hire questions, reducing anxiety and maintaining momentum throughout the onboarding process. Available around the clock, these bots ensure that support is accessible regardless of time zones or work schedules. FAQ automation streamlines information delivery by efficiently handling repetitive questions about policies, procedures, and benefits, freeing human resources staff to focus on more complex, relationship-building activities. Guided onboarding tasks provide step-by-step assistance through complex processes like benefits enrollment, system setup, and initial training modules, ensuring consistency and completeness while maintaining engagement through interactive dialogue.

Creating effective learning interactions requires careful attention to design principles that prioritize user experience and learning outcomes. Conversational design principles ensure that bot communications feel natural and engaging, using appropriate tone, clear language, and logical conversation flows that mirror human interaction patterns while maintaining professional standards. Personalized interactions adapt to individual communication styles and preferences, remembering previous conversations and building upon established rapport to create continuity throughout the onboarding journey. Timely and relevant content delivery ensures that information is presented precisely when needed, with contextually appropriate suggestions and proactive guidance that anticipates common challenges and provides solutions before problems arise, maintaining learner interest and supporting effective knowledge retention.

Successful interactive learning bot deployments demonstrate measurable improvements in onboarding outcomes and user satisfaction. Enhanced onboarding completion rates result from consistent guidance and support that prevents new hires from getting stuck or abandoning processes, with many organizations reporting completion rate improvements of 20-35%. Higher user satisfaction reflects the convenience and accessibility of instant support, with new hires appreciating the ability to get immediate answers without waiting for human assistance or feeling embarrassed about asking basic questions. Scalable interactive support enables organizations to maintain high-quality onboarding experiences even during periods of rapid hiring growth, providing consistent support quality that doesn't diminish with increased volume while reducing the burden on human resources teams.

Real-time feedback loops create dynamic onboarding experiences that continuously improve based on new hire input and engagement data. This section explores the mechanisms for collecting immediate feedback, analyzing responses, and implementing rapid improvements that enhance the onboarding experience for current and future employees. We'll examine how continuous feedback creates more responsive and effective onboarding programs.

Effective feedback collection requires diverse mechanisms that capture comprehensive insights without overwhelming new hires. Surveys for feedback provide detailed qualitative and quantitative data about specific aspects of the onboarding experience, enabling deep analysis of strengths and improvement opportunities while allowing new hires to share detailed suggestions and concerns. Pulse polls offer quick, frequent touchpoints that monitor satisfaction and engagement with minimal time investment, providing early warning indicators of potential issues before they impact retention or performance. Embedded feedback tools integrate seamlessly within onboarding platforms, enabling real-time collection of impressions and reactions without disrupting the learning flow, creating natural opportunities for input that feel organic rather than burdensome.

Translating feedback into actionable improvements requires systematic approaches to analysis and implementation. The importance of continuous feedback lies in its ability to identify gaps between intended and actual experiences, revealing disconnects that might otherwise go unnoticed until they impact retention or performance. Regular feedback analysis enables organizations to spot trends and patterns across multiple new hire cohorts, identifying systemic issues that require structural changes rather than individual interventions. Adjusting onboarding materials based on feedback ensures that content remains current, relevant, and effective, while methodology updates address process inefficiencies and incorporate best practices discovered through new hire experiences, creating evolutionary improvement that keeps onboarding aligned with employee needs and organizational goals.

Real-time feedback significantly impacts employee engagement by creating responsive environments that demonstrate organizational commitment to employee success. Immediate feedback benefits include making new hires feel heard and valued from their first day, as their input directly influences their experience and that of future employees. This responsiveness builds trust and demonstrates that the organization genuinely cares about employee perspectives and continuous improvement. Enhanced engagement emerges from feedback loops that create ongoing dialogue between new hires and the organization, fostering a culture of open communication and continuous learning. This engagement extends beyond onboarding, establishing patterns of feedback and improvement that support long-term employee development, retention, and organizational success through collaborative problem-solving and shared ownership of the employee experience.

Chapter 8: Personalized Onboarding Experiences

In conclusion, the four pillars of innovative onboarding create a comprehensive framework for new hire success. Personalized onboarding approaches leverage AI and data analytics to tailor experiences to individual employee needs, ensuring relevance and engagement from day one. Interactive bots provide scalable, real-time communication and support that enhances accessibility and consistency across all new hire experiences. Real-time feedback mechanisms enable continuous improvement and adaptability, creating responsive programs that evolve based on actual user experiences rather than assumptions. Together, these elements establish the foundation for long-term success, setting the stage for enhanced employee engagement, improved retention, and organizational growth through effective talent integration and development.

Chapter 9: Continuous Listening and Sentiment Analysis

Let's begin with continuous listening and sentiment analysis - the foundation of modern employee engagement strategies. This approach represents a fundamental shift from periodic check-ins to real-time understanding of your workforce. We'll explore why this matters, what technologies enable it, and how organizations are successfully implementing these systems to create more responsive, engaged workplace cultures.

Real-time employee feedback transforms how organizations respond to workforce needs. Prompt response capabilities mean you can address concerns before they escalate into larger problems. This proactive approach fosters deeper engagement by showing employees their voices matter and are heard immediately. Most importantly, continuous listening builds trust between leadership and staff, creating a culture where honest communication flows freely and workplace satisfaction improves naturally through responsive leadership.

Three core technologies power effective continuous listening systems. Natural Language Processing analyzes the nuanced meaning behind employee communications, capturing sentiment that traditional surveys might miss. Machine Learning algorithms continuously improve their understanding of your workforce patterns, becoming more accurate over time. Real-time Analytics provide immediate insights, enabling managers to respond quickly to emerging trends. Together, these technologies create intelligent systems that understand both what employees say and what they mean.

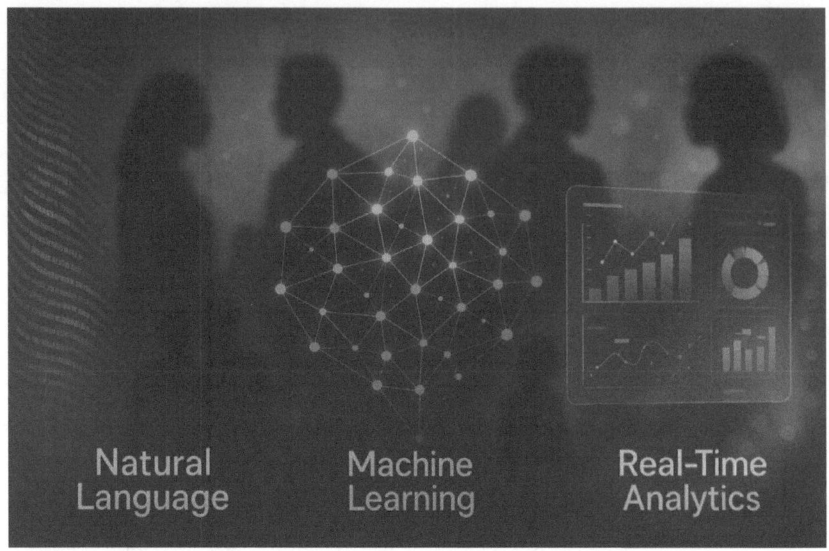

Modern sentiment analysis employs three primary methodologies, each with distinct advantages. Lexicon-based approaches use predefined word dictionaries with sentiment scores, providing quick baseline analysis. Supervised Machine Learning models learn from labeled training data to classify sentiments with increasing accuracy. Hybrid techniques combine both approaches, leveraging the speed of lexicon methods with the sophistication of machine learning. The best implementations use hybrid approaches to maximize both accuracy and efficiency.

Moving to our second major topic: pulse surveys enhanced with AI interpretation. While traditional surveys provide valuable snapshots, AI-powered pulse surveys offer continuous, intelligent analysis of employee sentiment. We'll explore how to design effective pulse surveys and leverage artificial intelligence to extract deeper insights from employee responses than ever before possible.

Effective pulse surveys require careful design principles. Keep them concise and relevant - focus on key engagement drivers rather than exhaustive questionnaires that cause survey fatigue. Conduct them frequently and timely to capture real-time insights and maintain organizational responsiveness. Ensure user-friendly completion through intuitive interfaces and mobile optimization. When employees find surveys easy and meaningful to complete, participation rates increase significantly, providing richer data for organizational decision-making.

Artificial intelligence revolutionizes survey analysis capabilities. Rapid processing allows organizations to analyze thousands of open-ended responses in minutes rather than weeks. AI excels at sentiment trend identification, detecting subtle shifts in employee mood across departments and timeframes. Perhaps most valuable is emerging issue detection - AI can identify new concerns before they become widespread problems, enabling proactive interventions that prevent larger organizational challenges from developing.

Extracting actionable insights requires strategic data correlation and prioritization. Link employee sentiment data directly to business outcomes like productivity, retention, and customer satisfaction to identify key performance drivers. Use this analysis to prioritize improvement areas, focusing resources on issues with the greatest organizational impact. This data-driven approach ensures interventions address root causes rather than symptoms, creating sustainable improvements in employee experience and business performance.

Our third focus area examines monitoring employee communication tools. As workplace communication increasingly moves digital, these platforms generate valuable sentiment and engagement data. We'll discuss various communication platforms, address critical ethical considerations, and explore how to analyze communication data while maintaining employee trust and legal compliance.

Modern workplaces utilize diverse communication platforms, each offering unique insights. Email remains fundamental for formal, structured communications and important announcements. Instant messaging platforms capture informal conversations and real-time reactions to organizational changes. Collaboration tools integrate communication with work processes, revealing team dynamics and project sentiment. Internal social networks foster community building and cultural engagement. Each platform provides different perspectives on employee experience and organizational health.

Ethical implementation requires careful attention to three fundamental principles. Transparency in monitoring means clearly communicating what data is collected and how it's used, maintaining employee trust through openness. Obtaining informed employee consent ensures legal compliance and respects individual privacy rights. Strict adherence to data protection laws like GDPR and local regulations protects both employees and organizations from legal risks. These ethical foundations are non-negotiable for sustainable communication monitoring programs.

Communication data analysis reveals three critical organizational insights. Sentiment shift detection helps identify changing employee moods and reactions to organizational events or decisions. Communication bottleneck identification reveals where information flow breaks down, hampering productivity and employee satisfaction. Engagement measurement assesses how actively employees participate in organizational communications, indicating overall connection to company culture. These insights enable targeted improvements to communication effectiveness and employee experience.

Finally, we'll explore predictive models for attrition - perhaps the most impactful application of employee analytics. By identifying employees at risk of leaving before they make that decision, organizations can implement targeted retention strategies. We'll examine data sources, model building approaches, and practical applications for improving workforce stability.

Effective attrition prediction draws from four key data sources. HR records provide foundational employee information including tenure, role changes, and compensation history. Engagement surveys capture satisfaction levels and commitment indicators that correlate with retention. Communication patterns reveal changing interaction levels that often precede departures. Performance metrics indicate productivity trends and career trajectory satisfaction. Combining these data sources creates comprehensive employee profiles that enable accurate attrition risk assessment.

Building robust predictive models requires systematic approaches. Machine Learning algorithms identify complex patterns in employee data that human analysis might miss, creating sophisticated prediction capabilities. Historical data validation ensures models accurately reflect past attrition patterns and can reliably predict future risks. Continuous model refinement incorporates new data and changing organizational dynamics, maintaining prediction accuracy over time. This iterative approach creates increasingly sophisticated tools for workforce planning and retention strategy.

Predictive insights enable three powerful retention strategies. Attrition predictions identify specific employees at risk, allowing managers to initiate proactive conversations and interventions. Tailored development programs address individual career growth needs and skill development interests that increase job satisfaction. Workload adjustments and enhanced support systems reduce burnout and improve work-life balance. These targeted approaches demonstrate organizational investment in employee success, significantly improving retention rates and reducing costly turnover.

In conclusion, continuous listening transforms organizational responsiveness by capturing real-time employee feedback and enabling proactive engagement strategies. Sentiment analysis provides deeper understanding of employee emotions and workplace experience, moving beyond surface-level metrics to genuine insights. When combined effectively, these approaches create enhanced employee experiences that foster engagement, satisfaction, and retention. The future of employee experience lies in intelligent, responsive systems that truly understand and support workforce needs.

Chapter 10: Engagement and Wellbeing Tools

Employee engagement platforms serve as the foundation of modern workplace wellness initiatives. These comprehensive systems facilitate seamless communication across teams, breaking down silos and enhancing collaboration through transparent information sharing. The feedback mechanisms built into these platforms enable continuous improvement by gathering real-time insights from employees at all levels. Recognition and rewards features motivate staff by acknowledging contributions promptly and meaningfully, creating positive reinforcement loops. Most importantly, these platforms foster a supportive workplace culture that directly impacts satisfaction and retention rates. When employees feel heard, valued, and connected, they're more likely to remain engaged and productive.

Technology has become the backbone of modern workplace wellbeing strategies, offering scalable solutions to complex human challenges. Digital health resources provide employees with 24/7 access to mental and physical health support through sophisticated apps and online platforms. Monitoring tools, including wearable devices and software applications, track key health metrics like stress levels, activity, and sleep patterns, enabling proactive wellbeing management. Continuous employee support is delivered through virtual coaching sessions, AI-powered chatbots, and online wellness communities that create ongoing engagement. This technological approach democratizes wellbeing support, making resources available regardless of location or schedule constraints, particularly valuable in today's hybrid work environments.

The business case for integrating wellbeing tools is compelling across multiple organizational metrics. Improved productivity emerges as employees maintain better focus and energy levels through targeted wellness interventions. Reduced absenteeism becomes evident as comprehensive wellbeing support addresses health issues before they require time off work. Enhanced morale creates a positive feedback loop where satisfied employees contribute to better workplace atmosphere and team dynamics. Stronger retention rates result from employees feeling valued and supported, reducing costly turnover and preserving institutional knowledge. Organizations investing in wellbeing tools consistently report measurable improvements in these areas, demonstrating clear return on investment while creating healthier work environments.

AI wellbeing applications represent a sophisticated evolution in employee health support, offering personalized interventions at scale.

Mood tracking capabilities monitor and record emotional patterns over time, providing valuable insights into mental health trends and potential triggers. Stress analysis features use advanced algorithms to assess stress levels and automatically provide personalized coping strategies tailored to individual needs and preferences. Virtual coaching offers interactive, real-time guidance that adapts to user responses and learning patterns, creating truly personalized wellness journeys. Health recommendations emerge from comprehensive data analysis, providing proactive suggestions that promote overall employee wellbeing. These features work synergistically to create comprehensive support systems that evolve with user needs.

The power of AI wellbeing applications lies in their ability to create truly personalized user experiences that adapt to individual preferences and behaviors. Adaptive user experiences dynamically adjust interfaces, content delivery, and interaction styles based on user preferences and past engagement patterns, ensuring maximum relevance and usability. Tailored content delivery ensures that each user receives information, recommendations, and interventions specifically relevant to their unique circumstances and goals. Behavior-based interventions analyze user patterns to provide targeted support at optimal moments, maximizing impact and engagement. This personalization creates more effective interventions because they feel natural and relevant to each user, increasing adoption rates and long-term engagement with wellbeing programs.

Leading AI wellbeing applications demonstrate the practical implementation of these advanced technologies in real workplace settings. AI mental health support platforms offer accessible, stigma-free mental health assistance through sophisticated chatbots and digital therapy tools that provide immediate support when needed. Wellness guidance platforms leverage AI to deliver personalized meditation, stress management, and mindfulness practices that adapt to individual schedules and preferences. Integration in workplaces shows how these applications seamlessly connect with existing HR systems and employee portals, creating comprehensive wellness ecosystems. These applications represent the cutting edge of workplace wellness technology, providing scalable solutions that can support thousands of employees while maintaining personalization and effectiveness.

Understanding burnout is crucial for effective workplace wellness strategies, as this pervasive issue affects both individual wellbeing and organizational performance. Burnout manifests as emotional, physical, and mental exhaustion resulting from prolonged exposure to workplace stress and overwhelming demands. The impact on productivity is severe, with burned-out employees showing decreased performance, creativity, and problem-solving abilities. The effect on overall wellbeing extends beyond work, affecting personal relationships, physical health, and mental state. Early identification and intervention are essential because burnout symptoms often develop gradually and can become deeply entrenched if not addressed. Organizations that understand burnout's multifaceted nature can implement more effective prevention and intervention strategies.

AI algorithms excel at identifying burnout risks through sophisticated analysis of multiple data streams and behavioral indicators. Behavioral data analysis examines patterns in work habits, communication frequency, task completion rates, and digital engagement to identify subtle changes that may indicate emerging burnout. Communication pattern detection monitors email frequency, response times, meeting participation, and collaboration patterns to spot concerning trends before they become critical. Physiological metrics monitoring integrates data from wearable devices and health apps to analyze sleep quality, heart rate variability, and activity levels. These algorithms can detect early warning signs that might be missed by traditional assessment methods, enabling proactive interventions.

Integration of burnout detection into organizational processes creates systematic approaches to employee wellbeing that complement existing HR frameworks. Burnout detection tools provide early warning systems that alert managers and HR professionals to at-risk employees, enabling timely intervention and support before issues escalate. Integration with HR systems streamlines the monitoring process by automatically connecting burnout risk assessments with employee records, performance data, and wellness program participation. Reducing employee turnover becomes achievable through proactive burnout management that addresses issues before employees reach the point of resignation. This systematic approach transforms reactive wellness strategies into proactive ones, creating more sustainable and effective workplace wellness programs.

Personalized wellness plans represent the pinnacle of AI-driven workplace wellness, combining multiple data sources to create truly individualized health strategies. Integration of health data involves collecting and analyzing diverse metrics from wearables, health assessments, productivity data, and employee surveys to create comprehensive wellness profiles. Consideration of preferences ensures that recommendations align with individual lifestyles, schedules, cultural backgrounds, and personal goals, increasing the likelihood of successful adoption. AI-driven wellness strategies analyze all available data to generate customized plans that address specific health needs, risk factors, and improvement opportunities. This personalized approach acknowledges that effective wellness interventions must be tailored to individual circumstances rather than applying one-size-fits-all solutions.

AI's role in analyzing wellbeing data transforms vast amounts of information into actionable insights that drive effective wellness strategies. Data processing efficiency allows AI systems to handle complex datasets from multiple sources simultaneously, identifying correlations and trends that would be impossible to detect manually. Pattern recognition capabilities identify meaningful relationships in wellbeing data, such as connections between stress levels, sleep quality, and productivity metrics. Wellness recommendations emerge from comprehensive analysis, providing evidence-based suggestions that are both personalized and grounded in data science. This analytical capability ensures that wellness interventions are based on objective evidence rather than assumptions, leading to more effective outcomes and better resource allocation.

Measuring the effectiveness of tailored recommendations requires robust evaluation systems that track outcomes and enable continuous improvement. Continuous data collection ensures accurate tracking of wellness plan outcomes over time, providing the feedback necessary to assess intervention effectiveness. Impact evaluation involves analyzing collected data to determine whether tailored recommendations are achieving desired health outcomes and behavioral changes. Guiding improvements uses evaluation feedback to refine and enhance wellness plans, creating iterative improvement cycles that optimize effectiveness. This measurement approach ensures that wellness programs remain evidence-based and continue evolving to meet changing employee needs and organizational goals.

In conclusion, AI-driven workplace wellness represents a transformative approach to employee health and organizational success. AI enhances employee wellbeing by providing sophisticated, personalized interventions that address individual needs while scaling across large organizations. Boosting productivity occurs naturally as healthier, more engaged employees perform better and contribute more effectively to organizational goals. Fostering supportive workplaces becomes achievable through technology that creates positive, data-driven wellness cultures. Organizations implementing comprehensive AI wellness strategies position themselves as employers of choice while building more resilient, productive workforces. The future of workplace wellness lies in these intelligent, adaptive systems that support both individual flourishing and organizational success.

Chapter 11: Personalized Learning Pathways

Personalized learning represents a fundamental shift from traditional educational models. Unlike standardized approaches, personalized learning adapts educational content to meet each individual learner's unique needs, learning style, and preferences. This approach recognizes that learners have different backgrounds, goals, and ways of processing information. By allowing flexible pacing, learners can progress at their own speed, spending more time on challenging concepts while accelerating through familiar material. Research consistently shows that personalization increases student motivation and leads to better academic achievement.

The benefits of individualized pathways extend far beyond traditional academic metrics. Enhanced motivation occurs when learners engage with content that resonates with their interests and connects to their personal goals. This relevance factor is crucial for adult learners and professional development contexts. Improved knowledge retention results from the brain's ability to better encode information when it's presented in familiar contexts and at appropriate difficulty levels. Most importantly, personalized pathways foster skill mastery and encourage lifelong learning habits essential for navigating evolving career landscapes.

Several technological advancements have made large-scale personalization possible. Learning Management Systems provide the infrastructure for organizing and delivering customized educational content efficiently across diverse learner populations. Data analytics transforms raw learning data into actionable insights, enabling real-time optimization of learner experiences. AI-powered learning platforms represent the cutting edge, using machine learning algorithms to adapt content dynamically based on learner behavior patterns, performance data, and predictive models that anticipate individual needs and preferences.

Understanding current and future skills requirements begins with comprehensive labor market analysis. By examining trends across industries, we can identify skills currently in high demand and anticipate future workforce needs. Industry forecasts provide crucial insights into how sector growth, technological advancement, and economic shifts will influence skill requirements. Emerging technologies like artificial intelligence, automation, and blockchain don't just create new job categories—they fundamentally alter existing roles, creating an urgent need for organizations and individuals to adapt through strategic upskilling and reskilling initiatives.

Effective skills assessment requires multiple evaluation methods to create a comprehensive competency profile. Self-assessments provide valuable insights into learners' perceived strengths and development areas, while peer reviews offer external perspectives on collaborative and interpersonal skills. Formal evaluations through standardized tests, practical demonstrations, and portfolio reviews provide objective measures of technical competencies. By identifying individual strengths, we can build personalized learning pathways that leverage existing capabilities while addressing specific development needs through targeted interventions and growth opportunities.

Chapter 11: Personalized Learning Pathways

Modern skills gap analysis relies on sophisticated tools and methodologies. Digital platforms centralize workforce data, enabling comprehensive analysis of skills inventories, performance metrics, and learning outcomes. AI-driven analytics process complex datasets to identify patterns, predict skill gaps, and recommend personalized development paths based on individual profiles and organizational needs. Competency frameworks provide structured approaches to defining required skills at different levels, enabling systematic evaluation and creating clear roadmaps for career advancement and professional development within specific industries or roles.

Adaptive learning technology represents a paradigm shift in educational delivery. These systems employ sophisticated algorithms that continuously analyze learner data—including response times, accuracy rates, learning preferences, and engagement patterns—to tailor educational content in real-time. Personalized content delivery means each learner receives materials optimized for their current knowledge level, learning style, and goals. The result is an optimized learning experience that maximizes efficiency and effectiveness by continuously adjusting to learner needs, eliminating the frustration of content that's too easy or overwhelming.

Adaptive systems personalize learning through four key mechanisms. Continuous learner analysis monitors every interaction, from correct answers to hesitation patterns, building detailed profiles of learner behavior and preferences. Customized resource recommendation suggests specific materials, exercises, and pathways based on individual performance and learning goals. Dynamic difficulty adjustment ensures learners are appropriately challenged without becoming overwhelmed or bored. Variable lesson pacing adapts to individual learning speeds, allowing thorough mastery of concepts before advancing while preventing high-achievers from becoming disengaged.

Real-world implementations of adaptive learning platforms demonstrate significant measurable outcomes. Enhanced learner engagement results from content that continuously adapts to maintain optimal challenge levels and relevance. Students report higher satisfaction and demonstrate increased participation when learning experiences feel personally meaningful. Improved knowledge retention occurs because adaptive systems reinforce concepts at optimal intervals and present information in multiple formats suited to individual learning preferences. Skill development accelerates through targeted feedback and practice opportunities that address specific competency gaps identified through continuous assessment.

Artificial intelligence is transforming career guidance through three primary capabilities. Data analytics enables AI systems to identify emerging career trends and opportunities by processing vast amounts of labor market data, salary information, and industry projections. Machine learning algorithms analyze individual profiles—including skills, experience, interests, and goals—to deliver highly personalized career advice and recommendations. Natural Language Processing allows AI systems to understand and interpret complex user queries, enabling natural, conversational interactions that make career guidance more accessible and effective for diverse user populations.

Personalized coaching powered by data analytics creates unprecedented opportunities for career development. AI skill assessment platforms evaluate individual competencies with greater accuracy and objectivity than traditional methods, identifying both strengths and development opportunities. Customized coaching programs combine individual preferences with market trends and industry demands to optimize career growth strategies. Advanced opportunity identification systems analyze market data to suggest career paths, job opportunities, and skill development priorities aligned with individual goals while considering broader economic trends and industry evolution.

The future of AI-driven career coaching promises more proactive and immersive experiences, with systems that anticipate user needs and provide personalized guidance before challenges arise. However, this advancement raises critical ethical considerations. Data privacy concerns require robust protections for sensitive career and personal information. Addressing bias and ensuring fairness in AI algorithms is essential to provide equitable career coaching across diverse populations. Transparency in AI decision-making processes fosters user trust and enables ethical accountability, ensuring that automated career guidance serves human interests rather than perpetuating existing inequalities.

Modern technology and AI have revolutionized education by enabling truly personalized learning pathways tailored to individual needs and career goals. Adaptive systems leverage data-driven insights to guide learners through rapidly evolving skill demands, ensuring relevant and effective development experiences. These tailored learning pathways support continuous career advancement, helping learners reach their full potential in an increasingly dynamic professional landscape. As we move forward, the integration of personalized learning, skills gap analysis, adaptive systems, and AI-driven career coaching will continue to transform how we approach education and professional development.

Chapter 12: AI-Driven Performance Management

AI-driven performance management represents a fundamental shift from traditional approaches. AI integration enables organizations to analyze vast amounts of employee data, transforming raw information into actionable insights that improve management decisions and strategic planning. This enhanced approach delivers more personalized, faster, and smarter management strategies that directly impact performance outcomes. By leveraging machine learning and predictive analytics, organizations can move beyond subjective assessments to create data-driven performance optimization systems that benefit both employees and the business.

The key differences between AI-driven and traditional performance management are substantial and transformative. Continuous performance monitoring replaces periodic reviews, providing ongoing insights rather than annual snapshots. Data-backed insights eliminate subjective opinions, ensuring decisions are grounded in objective analysis. Automation and predictive analytics enable systems to process information and predict future trends automatically. These differences create a more responsive, accurate, and effective performance management ecosystem that adapts to changing business needs and employee development requirements.

Adopting AI in performance management brings significant benefits alongside important challenges that organizations must address. Efficiency and accuracy gains improve data analysis and decision-making processes, while enhanced employee engagement results from personalized feedback and performance tracking. However, data privacy concerns require careful attention to sensitive employee information and compliance requirements. Algorithm bias and change management present ongoing challenges as organizations integrate AI into existing systems, requiring thoughtful implementation strategies and continuous monitoring to ensure ethical and effective deployment.

Real-time data collection and analysis form the foundation of continuous performance tracking. AI tools gather information instantly from multiple sources, providing up-to-date insights at any moment. This instant data gathering enables performance trend detection, allowing organizations to identify patterns early and support quick decision-making. Perhaps most importantly, AI systems excel at issue identification, spotting potential problems before they escalate and allowing for prompt corrective actions that minimize risks and maintain optimal performance levels across teams.

Automated goal setting and progress monitoring represent a significant advancement in performance management. AI-driven goal definition helps establish realistic objectives tailored to individual and team performance metrics and capabilities, ensuring targets are both challenging and achievable. Continuous progress monitoring tracks advancement in real-time, identifying areas needing improvement as they emerge. The system provides actionable feedback delivery, offering timely insights that help employees adjust their efforts and improve outcomes, creating a dynamic cycle of continuous improvement and professional development.

Enhancing employee engagement through ongoing insights creates a more motivated and committed workforce. Personalized employee insights help individuals understand their unique strengths and areas for growth, directly boosting engagement levels. Motivation increases through tailored recommendations that guide professional development pathways effectively, showing employees clear paths for advancement. Most importantly, this approach empowers employees to take ownership of their development, enhancing commitment and job satisfaction. When employees feel supported by data-driven insights, they become more invested in their roles and organizational success.

AI-powered bots deliver instant feedback through sophisticated data analysis and immediate response systems. These bots analyze performance data continuously to accurately assess user interactions and outcomes, providing context-aware insights. The instant feedback mechanism ensures employees receive immediate, constructive guidance that encourages timely learning and skill development. This continuous support system helps users refine their abilities and enhance performance over time, creating a culture of ongoing improvement where feedback becomes a natural part of the work experience rather than an occasional formal process.

Automating recognition and reward systems ensures fairness and scalability in employee appreciation programs. AI identification of achievements accurately recognizes individual contributions across various environments and contexts, eliminating the risk of overlooking valuable work. The system ensures consistent and unbiased rewards, as automation removes human bias from recognition decisions, creating fair treatment for all employees. Perhaps most importantly, AI-driven systems provide scalable recognition processes that can efficiently manage rewards across large organizations, ensuring no achievements go unnoticed regardless of company size or geographic distribution.

Improving team dynamics and morale through timely support creates a more collaborative and positive work environment. Timely feedback importance cannot be overstated, as providing input when it matters builds trust and helps team members improve effectively. Recognition boosts morale significantly, as acknowledging achievements motivates employees and fosters a positive workplace atmosphere where contributions are valued. This timely support enhances collaboration by encouraging better teamwork and strengthening overall team dynamics, creating an environment where employees feel supported, appreciated, and motivated to contribute their best efforts.

Reducing bias through AI algorithms represents a critical advancement in fair performance evaluation. Objective data analysis enables AI systems to minimize subjective human judgment and personal biases that can skew traditional performance assessments. Standardized evaluation criteria ensure that AI applies consistent standards across all evaluations, creating fair and unbiased assessments regardless of manager preferences or unconscious bias. This systematic approach to performance evaluation builds trust in the appraisal process and ensures that all employees are evaluated using the same objective standards and criteria.

Ensuring transparency in evaluation metrics builds employee confidence and trust in the performance management system. Data-driven metrics utilize clear, measurable data points to evaluate employee performance objectively and transparently, eliminating guesswork and subjective interpretation. Employee understanding improves significantly when transparent metrics help individuals comprehend exactly how their work is assessed, fostering both trust and motivation. When employees understand the evaluation criteria and can see how their performance is measured, they become more engaged in the improvement process and more confident in the fairness of their assessments.

Building trust with data-driven appraisals creates a foundation for effective performance management and employee satisfaction. Fairness in appraisals emerges naturally from data-driven approaches that reduce bias and provide objective performance assessments based on measurable outcomes. Consistency in evaluations ensures reliable appraisal outcomes across all employees and teams, eliminating favoritism and subjective variations. This approach enhances engagement and retention, as trust built through data-driven fairness significantly boosts employee satisfaction and their commitment to remaining with the organization long-term.

In conclusion, AI-driven performance management transforms organizations through four key mechanisms. Continuous performance tracking enables real-time monitoring to identify strengths and improvement areas promptly. Instant feedback mechanisms provide immediate guidance that fosters growth and engagement. Fair and objective appraisals promote equity by reducing bias and basing evaluations on data-driven insights. Together, these elements drive organizational success by supporting effective workforce strategies and boosting overall productivity, creating a competitive advantage through optimized human capital management and enhanced employee satisfaction.

Chapter 13: Succession Planning and Talent Mobility

Succession planning serves as your organization's insurance policy against leadership gaps, ensuring business continuity when key personnel leave or retire. It involves systematically identifying and developing internal talent to fill future leadership roles. Talent mobility, meanwhile, is the strategic movement of employees across different roles, departments, or geographic locations within your organization. This practice enhances employee potential by exposing them to diverse experiences and skill sets. Together, these practices create a robust pipeline of capable leaders while maximizing the value of your existing workforce investment.

These practices deliver four critical organizational benefits. Leadership continuity ensures smooth transitions during personnel changes, maintaining operational stability and strategic momentum. Risk reduction occurs when you minimize dependence on external hires for key positions, reducing recruitment costs and onboarding risks. Employee engagement naturally increases when workers see clear career advancement pathways within the organization, leading to higher retention rates. Strategic workforce alignment ensures your talent development efforts directly support business objectives, creating a competitive advantage through better skilled, more committed employees who understand your organizational culture and values.

Traditional succession planning faces three major challenges that limit its effectiveness. Subjective assessments rely heavily on personal opinions and informal observations, which can introduce bias and overlook hidden talent. Managers may favor employees who are more visible or similar to themselves, missing high-potential individuals in different departments or with different working styles. Limited data usage means decisions are made with incomplete information, lacking comprehensive performance metrics, skill assessments, and potential indicators. Informal processes without structured frameworks increase the risk of leadership gaps and missed development opportunities, leaving organizations vulnerable when key personnel unexpectedly leave.

Identifying high potential talent requires evaluating four key indicators that predict future leadership success. Performance metrics provide quantifiable evidence of an individual's current effectiveness, including productivity measures, goal achievement, and quality standards. Learning agility demonstrates an employee's ability to quickly acquire new skills and adapt knowledge to different situations, crucial for leadership in rapidly changing environments. Leadership capabilities encompass both formal and informal influence, including team collaboration, mentoring others, and driving positive change. Adaptability and engagement measure how well individuals cope with change while maintaining high motivation and commitment levels.

Data-driven approaches revolutionize talent identification through four key improvements over traditional methods. Advanced analytics leverage sophisticated algorithms to extract meaningful insights from diverse data sources, including performance records, 360-degree feedback, and behavioral assessments. Machine learning models continuously improve their predictive accuracy by identifying patterns in successful leadership transitions and career progressions. Bias reduction occurs when objective data replaces subjective opinions, creating fairer and more reliable talent identification processes. Improved accuracy results from combining multiple data points and predictive models, providing more comprehensive assessments of employee potential and readiness for advancement.

Early identification of high potential talent creates three significant advantages for leadership pipeline development. Tailored development programs can be customized to individual strengths and growth areas, accelerating leadership readiness through targeted training, mentoring, and stretch assignments. This personalized approach maximizes development investments and employee engagement. Reduced employee turnover occurs when organizations demonstrate commitment to employee growth and career advancement, creating stronger retention rates among high performers. Steady leadership pipeline ensures continuous availability of qualified internal candidates for key positions, reducing external recruitment costs and maintaining organizational knowledge and culture continuity.

Chapter 13: Succession Planning and Talent Mobility

Effective role matching employs three sophisticated techniques to optimize employee-position alignment. Skills and competency assessments provide comprehensive evaluations of employee capabilities against specific job requirements, using validated tools to measure technical skills, soft skills, and potential for role success. Behavioral profiling analyzes personality traits, work preferences, and behavioral patterns to match employees with roles that suit their natural tendencies and motivations. AI-driven recommendation systems leverage machine learning algorithms to analyze vast amounts of employee and role data, identifying optimal matches based on historical success patterns and predictive modeling, significantly improving placement accuracy.

Internal mobility strategies encourage employee growth through three primary mechanisms. Lateral moves provide opportunities to gain diverse skills and experience across different departments or functions, broadening employee perspectives and increasing organizational flexibility. These moves often reignite employee engagement and provide fresh challenges. Cross-functional assignments create collaborative opportunities across departments while expanding employee experience and understanding of business operations. Upskilling opportunities through training, certification, and development programs prepare employees for future roles while addressing current skill gaps, creating win-win scenarios for both employee advancement and organizational capability building.

Modern talent movement platforms provide three essential capabilities for effective internal mobility programs. Internal job matching systems use sophisticated algorithms to connect employees with suitable opportunities based on skills, preferences, career goals, and organizational needs, making internal opportunities more visible and accessible. Career path tracking tools help employees visualize potential advancement routes while providing managers with insights into employee aspirations and development needs. Managerial insights dashboards offer analytics on talent movement patterns, success rates, and organizational capability gaps, enabling data-driven decisions about workforce development and resource allocation strategies.

AI integration transforms workforce planning through four key capabilities that enhance strategic decision-making. Workforce data analysis processes large, complex datasets to uncover trends, patterns, and insights that would be impossible to identify manually, providing deeper understanding of workforce dynamics. Talent needs forecasting uses predictive algorithms to anticipate future skill requirements and headcount needs based on business plans, market trends, and historical data. Skill gap identification systematically compares current workforce capabilities against future requirements, highlighting specific areas needing attention. Proactive resource allocation enables strategic workforce planning based on predictive insights rather than reactive responses to immediate needs.

Predictive analytics enhances strategic decision-making through three critical applications in talent management. Succession strategy insights help identify potential leaders earlier and guide targeted development investments, improving leadership pipeline strength and reducing succession risks. Risk mitigation capabilities allow organizations to anticipate workforce challenges such as retirement waves, skill obsolescence, or market talent shortages, enabling proactive countermeasures. Optimizing talent deployment ensures the right people are in the right roles at the right time, maximizing both individual performance and organizational effectiveness through data-driven placement and development decisions that align with business objectives.

Three emerging trends will shape the future of AI-driven talent management over the next decade. Augmented intelligence combines AI capabilities with human judgment to enhance personalized career development, providing employees with AI-powered insights while maintaining human oversight for nuanced decisions. Real-time workforce analytics deliver continuous insights into workforce performance, engagement, and trends, enabling more agile management responses to changing conditions. Automation in talent operations streamlines routine tasks such as resume screening, initial candidate matching, and basic performance tracking, freeing HR professionals to focus on strategic initiatives and complex relationship management activities.

Predictive analytics enables organizations to anticipate workforce needs and prepare strategically for future challenges, moving from reactive to proactive talent management. AI revolutionizes talent mobility by identifying optimal role fits and creating efficient career pathing, reducing time-to-productivity and increasing employee satisfaction. The strategic advantage lies in creating an agile, well-prepared workforce that can adapt to changing business needs while maintaining high performance levels. Organizations implementing these approaches gain competitive advantages through better talent utilization, reduced turnover, and stronger leadership pipelines.

Chapter 14: Ethical Considerations in HR AI

Fairness in HR AI systems encompasses three critical dimensions that organizations must prioritize. Equal treatment and opportunity requires that all candidates and employees receive consistent, unbiased evaluation regardless of their background or characteristics. This foundation prevents systemic discrimination and ensures merit-based decisions. Preventing bias and favoritism involves implementing safeguards against both conscious and unconscious preferences that could skew AI recommendations. Finally, promoting diversity and inclusion means actively designing AI systems that support workplace equity by providing equal opportunities across all demographic groups, thereby strengthening organizational culture and performance through diverse perspectives and experiences.

Transparency in AI-powered HR decisions serves multiple crucial functions for organizational success. Understanding AI decisions means stakeholders can comprehend how algorithms process information and reach conclusions, enabling informed decision-making and system improvements. Building trust through transparency establishes credibility with employees, candidates, and management by demonstrating that AI systems operate fairly and consistently. When people understand the decision-making process, they're more likely to accept outcomes and engage positively with the system. Ensuring ethical standards becomes possible when AI processes are open to scrutiny, allowing organizations to identify potential issues, maintain compliance with regulations, and uphold their values consistently.

Explainability represents the practical implementation of transparency in AI systems, focusing on clear communication of decision-making processes. The definition encompasses the ability to articulate in understandable terms how AI systems analyze data and reach specific conclusions in HR contexts. For HR professionals, explainability provides the knowledge needed to validate AI recommendations, make informed overrides when necessary, and communicate decisions effectively to stakeholders. Most importantly, explainability empowers individuals affected by AI decisions by giving them the right to understand and challenge outcomes. This capability is essential for maintaining employee trust, meeting legal requirements, and ensuring that automated systems remain accountable to human judgment and organizational values.

Identifying and mitigating biases in HR datasets requires a systematic approach to data analysis and algorithmic design. Bias identification involves comprehensive analysis of data sources, historical patterns, and algorithmic outputs to detect discriminatory trends that could affect decision-making. This process includes examining representation across demographic groups, identifying correlations between protected characteristics and outcomes, and testing for disparate impact. Bias mitigation techniques encompass various technical approaches including data balancing to ensure representative samples, fairness-aware algorithms that explicitly account for equity considerations, preprocessing methods to remove biased features, and postprocessing adjustments to ensure equitable outcomes across all groups while maintaining predictive accuracy.

Preventing discriminatory outcomes in recruitment and performance evaluation requires multiple complementary strategies working in concert. AI model auditing involves regular systematic reviews of algorithmic performance across different demographic groups, identifying potential biases and measuring fairness metrics to ensure equitable treatment. Fairness constraints are mathematical requirements built directly into AI algorithms that explicitly prevent discriminatory outcomes by optimizing for equity alongside accuracy. Diverse training data ensures algorithms learn from representative samples that reflect the full spectrum of qualified candidates and high-performing employees. Continuous monitoring establishes ongoing surveillance of AI system performance, enabling rapid detection and correction of emerging biases as organizational contexts and external factors evolve over time.

Legal and ethical guidelines provide the framework for responsible AI implementation in HR contexts, ensuring compliance and accountability. Legal compliance requires adherence to anti-discrimination laws, employment regulations, and data protection requirements that vary by jurisdiction but consistently prohibit unfair treatment based on protected characteristics. Organizations must understand applicable laws and design AI systems that meet or exceed these standards. Ethical accountability goes beyond legal minimums to establish responsible practices that reflect organizational values and societal expectations. Privacy respect involves protecting personal data through secure handling, limited collection, transparent usage policies, and giving individuals control over their information, creating trust and demonstrating commitment to ethical AI deployment.

Human oversight in AI systems presents both significant benefits and important limitations that organizations must carefully balance. Error detection capabilities allow humans to identify mistakes, edge cases, and contextual factors that automated systems might miss, providing essential quality control. Ethical judgment enables human reviewers to apply moral reasoning, consider broader implications, and ensure decisions align with organizational values and societal expectations that algorithms cannot fully capture. However, potential delays and bias represent limitations where human involvement may slow decision-making processes and introduce subjective preferences or inconsistencies. The key is designing oversight systems that maximize human strengths while minimizing these weaknesses through structured processes and clear guidelines.

Establishing effective review processes for AI-generated HR decisions requires structured approaches that ensure quality and accountability. Clear auditing protocols provide well-defined procedures for reviewing AI outputs, including specific criteria for evaluation, documentation requirements, and escalation procedures when issues arise. These protocols ensure consistency and thoroughness in the review process.

Multidisciplinary team involvement brings together HR professionals, data scientists, legal experts, and other stakeholders to provide comprehensive evaluation from multiple perspectives, enhancing decision quality and catching issues that single reviewers might miss. Feedback mechanisms create continuous improvement loops where review outcomes inform system refinements, training data updates, and algorithmic adjustments, ensuring AI systems evolve and improve over time.

Balancing automation with human judgment requires careful integration of technological efficiency with human wisdom and accountability. AI efficiency and human insight creates optimal outcomes by leveraging algorithms' ability to process large datasets consistently while incorporating human understanding of context, nuance, and exceptional circumstances that require flexible thinking. Ethical and data-driven decisions emerge when human oversight ensures that algorithmic recommendations align with moral principles and organizational values while still benefiting from comprehensive data analysis. Human responsibility in HR maintains the principle that final decisions rest with people who can be held accountable, understand broader implications, and exercise judgment that considers factors beyond algorithmic calculations, preserving the human element in employment decisions.

In conclusion, implementing ethical AI in HR systems requires attention to three interconnected principles that work together to create fair and effective processes. Fairness in AI demands that systems treat all employees and candidates equitably, eliminating discrimination and bias while promoting inclusive practices that benefit organizations and individuals alike. Transparency and accountability ensure that AI processes remain understandable and justifiable, building trust through clear communication and maintaining responsibility for decisions that affect people's careers and lives. Human oversight provides the essential element of judgment, ethics, and accountability that ensures AI systems serve human needs and organizational values rather than operating as opaque black boxes that remove human agency from critical employment decisions.

Chapter 15: Change Management and HR Transformation

Change management in HR requires a strategic approach built on three foundational pillars. Flexibility stands as the cornerstone, enabling organizations to adapt their strategies as needs evolve and new challenges emerge. Employee engagement represents the human element, ensuring that transformation initiatives gain buy-in from those most affected by the changes. When employees feel heard and involved, resistance diminishes and adoption accelerates. Alignment with organizational goals ensures that change efforts contribute meaningfully to broader business objectives. Without this alignment, transformation initiatives risk becoming isolated projects that fail to deliver sustainable value or lasting organizational improvement.

Three primary drivers are reshaping HR practices in the digital era. Technological advancements continue to revolutionize how HR operates, introducing automation capabilities and sophisticated data analytics that enhance decision-making processes. Modern workforce expectations have shifted dramatically, with employees demanding greater flexibility, inclusion, and continuous learning opportunities. This evolution requires HR to innovate their approaches and service delivery models. AI integration represents perhaps the most significant driver, transforming recruitment processes, performance management systems, and employee engagement strategies. Organizations that embrace these drivers position themselves for success, while those that resist face increasing competitive disadvantages.

Chapter 15: Change Management and HR Transformation

Leadership plays a pivotal role in navigating organizational transformation successfully. Effective leaders guide their teams through uncertainty, providing clear direction and maintaining momentum during challenging periods. They foster innovation cultures that encourage creative problem-solving and experimentation, recognizing that transformation requires new ways of thinking and working. Building resilience and promoting continuous learning becomes essential as AI-driven changes accelerate. Leaders must model adaptability themselves while supporting their teams through the emotional and practical challenges of transformation. Strong leadership creates the psychological safety necessary for employees to embrace change rather than resist it, ultimately determining the success or failure of transformation initiatives.

HR professionals must develop three essential AI competencies to remain effective in the digital age. Data literacy enables practitioners to analyze and interpret AI-generated insights, transforming raw information into actionable strategies. Without this skill, HR professionals cannot leverage the full potential of AI tools or make informed decisions based on algorithmic outputs. Understanding AI ethics ensures responsible implementation while protecting employee privacy and maintaining fairness in automated processes. This knowledge becomes critical as AI systems influence hiring, performance evaluation, and career development decisions. AI tools integration capability allows HR teams to streamline processes and improve efficiency, but requires understanding both technical capabilities and human implications of automated systems.

Developing effective learning and development programs requires a systematic approach to skill enhancement. Begin by conducting thorough skill gap analyses to identify specific areas where HR professionals need development. This assessment should examine both current capabilities and future requirements driven by AI adoption. Customize training programs to match the unique learning preferences and professional development needs of your HR teams, recognizing that one-size-fits-all approaches rarely succeed. Implement learning strategies that maximize upskilling effectiveness by combining theoretical knowledge with practical application opportunities. Consider blended learning approaches that include hands-on experience with AI tools, peer learning sessions, and ongoing mentorship to ensure skills are not only acquired but successfully applied in real workplace situations.

Measuring upskilling initiative effectiveness requires comprehensive evaluation approaches. Establish clear evaluation metrics that assess both skill acquisition and practical application in workplace scenarios. These metrics should include pre- and post-training assessments, on-the-job performance indicators, and long-term retention rates. Monitor the impact on HR capabilities by tracking improvements in process efficiency, decision-making quality, and service delivery. Look for measurable changes in how HR teams handle data analysis, implement AI tools, and navigate ethical considerations. Assess organizational transformation outcomes by examining how training initiatives contribute to broader cultural change, innovation adoption, and departmental adaptability. Regular evaluation ensures continuous improvement and demonstrates the value of upskilling investments to organizational stakeholders.

Effective stakeholder management begins with comprehensive mapping and understanding of key participants in the transformation process. Identify all individuals and groups who will be impacted by or can influence AI adoption initiatives, including employees, managers, executives, IT teams, and external partners. Understanding stakeholder concerns requires deep analysis of their unique perspectives, potential resistance points, and underlying motivations. Some may fear job displacement, others might worry about privacy implications, while leadership may focus on ROI and competitive advantages. Tailored communication strategies must address these specific concerns while fostering collaboration and buy-in. Successful stakeholder management prevents resistance from derailing transformation efforts and builds the coalition necessary for sustainable change implementation.

Crafting effective communication plans requires attention to three critical elements that ensure message success across diverse stakeholder groups. Clarity in messaging means developing communications that are easily understood regardless of technical background or organizational level. Avoid jargon and clearly explain both the rationale for change and practical implications for each audience. Consistency across channels builds trust and prevents confusion by ensuring that stakeholders receive the same core messages whether through emails, meetings, intranet posts, or training sessions. Timely information delivery keeps stakeholders informed and engaged throughout the transformation process. Delayed or sporadic communication creates uncertainty and anxiety, while proactive updates demonstrate organizational transparency and commitment to inclusive change management.

Digital tools significantly enhance stakeholder engagement capabilities throughout transformation initiatives. Enhanced communication platforms provide real-time, interactive channels that facilitate two-way dialogue between leadership and employees. These tools enable immediate clarification of questions, rapid dissemination of updates, and dynamic discussion forums that build community around change initiatives. Accessible feedback channels ensure that stakeholder input can be captured conveniently and continuously, rather than being limited to formal survey periods or scheduled meetings. Collaborative engagement tools enable shared decision-making through virtual workspaces, project management platforms, and digital collaboration environments. These technologies democratize participation in transformation planning and implementation, ensuring that diverse perspectives inform change strategies and increase overall stakeholder investment in successful outcomes.

Assessing AI adoption readiness requires evaluation across three critical dimensions of organizational capability. Technological readiness examines your current infrastructure's ability to support AI implementation, including hardware capacity, software compatibility, data management systems, and cybersecurity protocols. This assessment identifies technical gaps that must be addressed before successful AI deployment. Cultural readiness evaluates your organization's openness to technological change, willingness to embrace new working methods, and capacity for continuous learning. Cultural resistance often presents the greatest barrier to AI adoption success. Operational readiness measures how well existing processes and workflows can accommodate AI integration without disrupting essential business functions or compromising service quality.

Conducting thorough gap analysis requires systematic evaluation of current capabilities against desired future states. Understanding your current state involves comprehensive assessment of existing resources, skills, processes, and cultural attributes that will influence AI adoption success. This baseline assessment must be honest and detailed to ensure accurate planning. Defining the desired state requires clear articulation of future goals, required capabilities, and expected outcomes from AI integration. These targets should align with broader organizational strategy and stakeholder expectations. Identifying gaps involves analyzing disparities between current and desired states to prioritize improvement efforts. This analysis informs resource allocation decisions, training program development, and timeline planning for successful transformation implementation.

Action planning transforms gap analysis insights into concrete steps for addressing readiness deficiencies. Begin by recognizing and analyzing specific gaps that hinder transformation readiness, prioritizing those that pose the greatest risks to successful AI adoption. Develop concrete action plans with clear timelines, assigned responsibilities, and measurable outcomes that directly address identified readiness gaps. These plans should include resource requirements, risk mitigation strategies, and success criteria. Implement a structured transformation approach that ensures systematic progress while maintaining operational continuity. This methodical approach reduces the likelihood of overlooking critical preparation steps and provides clear milestones for tracking transformation progress. Regular plan reviews and adjustments ensure that action plans remain relevant as organizational needs evolve during the transformation journey.

Successful AI-driven HR transformation depends on four interconnected elements working in harmony. Strong leadership provides the vision, direction, and support necessary to guide organizations through complex change processes. Leaders must champion transformation initiatives while addressing employee concerns and resistance. Targeted upskilling ensures that employees develop the specific competencies needed to thrive in AI-enhanced work environments, focusing training efforts on areas with greatest impact potential. Effective communication maintains alignment among all stakeholders, preventing misunderstandings that could derail transformation efforts. Clear, consistent messaging builds trust and reduces anxiety about change. Comprehensive readiness assessment prepares organizations to confidently embrace AI-driven changes by identifying and addressing potential obstacles before they become significant problems.

Chapter 16: Measuring Impact and ROI

Successful AI implementation in HR requires careful measurement across four key performance areas. Recruitment efficiency focuses on how quickly and effectively AI tools fill open positions, reducing time-to-hire and improving candidate quality. Employee engagement metrics track satisfaction and participation levels, indicating AI's impact on workplace culture. Retention rates reveal AI's ability to predict and prevent turnover through data-driven insights. Process automation effectiveness measures how well AI streamlines administrative tasks and reduces manual workload.

Effective evaluation requires both quantitative and qualitative metrics. Quantitative measures include hard data like time-to-hire, cost savings, and productivity improvements that demonstrate clear financial impact. Qualitative insights capture employee satisfaction, perceived fairness, and cultural impact that numbers alone cannot reveal. The most comprehensive evaluation combines both approaches, providing a complete picture of AI's organizational impact. This dual approach ensures decisions are based on concrete data while considering human factors essential for long-term success.

Interpreting AI metrics requires contextual analysis within your organizational framework to ensure insights are relevant and actionable. Strategic optimization involves continuously refining AI tools based on performance data to enhance HR decision-making and operational efficiency. Successful implementation aligns AI-driven insights with broader organizational objectives, ensuring maximum impact on business goals. This strategic approach transforms raw data into meaningful guidance that drives competitive advantage and organizational growth through informed HR decisions.

Establishing baseline benchmarks begins with analyzing historical HR data to create performance baselines for measuring AI initiative success. Initial AI performance assessment involves evaluating early results to understand current capabilities and identify areas for improvement. Setting realistic targets based on historical performance and AI capabilities ensures achievable goals that drive meaningful progress. This foundation enables accurate measurement of AI impact and provides clear direction for optimization efforts throughout the implementation process.

Internal performance evaluation assesses AI-driven HR metrics within your organization to measure current effectiveness and identify improvement opportunities. External benchmarking compares performance against industry standards and competitor data, revealing market position and identifying best practices for adoption. This dual approach helps identify innovation opportunities by revealing gaps where AI-driven HR practices can provide competitive advantage. Combining internal and external perspectives ensures comprehensive understanding of performance relative to both organizational goals and market standards.

Continuous feedback collection involves regularly gathering input from users and stakeholders to identify improvement areas in AI and HR systems. Integrating feedback requires systematically incorporating collected insights to refine AI systems and optimize HR processes for better outcomes. Ongoing enhancement uses iterative improvements based on feedback loops to achieve continuous advancement of technology and workflows. This cyclical process ensures AI systems evolve with organizational needs and maintain effectiveness over time through responsive adaptation.

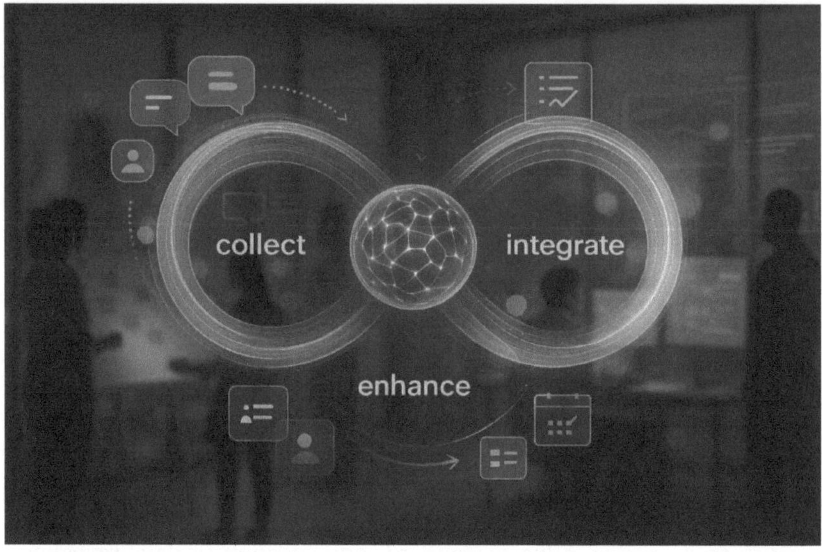

AI adoption goals must clearly define objectives for HR implementation, focusing on improving operational efficiency and enhancing employee satisfaction. These goals provide direction and measurable targets for success evaluation. Expected value outlines specific benefits anticipated from AI initiatives, including streamlined processes, enhanced employee experience, and improved decision-making capabilities. Clear objective definition ensures all stakeholders understand the purpose and potential impact of AI investments, creating alignment and realistic expectations for implementation outcomes.

Financial investment assessment evaluates total project costs against expected outcomes and benefits, providing clear ROI calculations for stakeholder review. Cost savings analysis identifies potential reductions in operational expenses through efficient resource allocation and elimination of unnecessary processes. Productivity gains measure improvements resulting from optimized resource planning and automated workflows. This comprehensive financial analysis demonstrates the business value of AI investments and supports informed decision-making about resource allocation and implementation priorities.

Highlighting strategic benefits emphasizes how AI projects align with long-term organizational goals and drive competitive advantage in the marketplace. Demonstrating ROI requires providing clear, data-driven evidence of financial returns to convince stakeholders of project value. Securing stakeholder support involves engaging key decision-makers through transparent communication, building trust, and obtaining necessary approvals for implementation. Effective communication strategies translate technical benefits into business language that resonates with various stakeholder groups and facilitates buy-in.

Metric selection importance emphasizes that choosing relevant, meaningful measurements is crucial for accurately assessing AI-driven HR initiatives' success and organizational impact. Effective benchmarking practices against industry standards help organizations track progress and ensure AI efforts align with organizational goals and market expectations. Strong business case development ensures AI investments provide measurable, meaningful organizational value that justifies resource allocation. These three pillars work together to create a comprehensive framework for successful AI implementation and sustained organizational benefit.

Printed by Libri Plureos GmbH in Hamburg, Germany